Frederick T Parson

Vital Magnetism

Its power over disease. A statement of the facts developed by men who have

employed this agent under various names, as animal magnetism, mesmerism,

hypnotism, etc., from the earliest times down to the present

Frederick T Parson

Vital Magnetism
Its power over disease. A statement of the facts developed by men who have employed this agent under various names, as animal magnetism, mesmerism, hypnotism, etc., from the earliest times down to the present

ISBN/EAN: 9783337240554

Printed in Europe, USA, Canada, Australia, Japan

Cover: Foto ©berggeist007 / pixelio.de

More available books at **www.hansebooks.com**

VITAL MAGNETISM:

Its Power Over Disease.

A STATEMENT OF THE FACTS DEVELOPED BY MEN WHO HAVE
EMPLOYED THIS AGENT UNDER VARIOUS NAMES, AS
ANIMAL MAGNETISM, MESMERISM, HYPNOTISM,
ETC., FROM THE EARLIEST TIMES
DOWN TO THE PRESENT.

BY

FREDERICK T. PARSON,

(*Magnetic Physician*).

"The essential point, where any question arises respecting
facts that are extraordinary and difficult to conceive, is not to
demonstrate how they exist, but to prove that they do exist."
NICOLE.

NEW YORK:
ADAMS, VICTOR & CO., PUBLISHERS,
98 WILLIAM STREET.
1877.

PUBLISHERS' NOTICE.

In presenting this volume to the public, the publishers bespeak for it the candid attention of the medical profession and of the press.

It will be found, we think, to present the subject of VITAL MAGNETISM in a manner befitting its importance to the human race as an agent of beneficence, while it must serve greatly to disabuse the minds of intelligent persons of impressions and prejudices fostered by the use—or rather *abuse*—of this mysterious power by charlatans.

The author's aim being to reach the discriminating reader, he has confined himself to FACTS whose authenticity is unquestioned, and to the experience of men of eminence in the medical and scientific world, whose words and work are conclusive as to the power, range and efficiency of VITAL MAGNETISM, both as a curative agent and as a subject meriting the most careful further observation to obtain a more determinate knowledge of its true nature and source.

Of the array of facts and observations here embodied by Dr. Parson, those relating to the case of Miss Harriet Martineau will command particular attention. This eminent woman had what may be termed a *marvelous experience*, which she narrated with a minuteness that offered invaluable data to the physician and investigator, but her editors and biographers have suppressed her own record of this wonderful recovery from an incurable disorder. Dr. Parson very properly reproduces the most essential portions of her singularly interesting narrative.

As stated in the announcement : " The volume is stuffed with facts, giving to theory and speculation upon the nature of the principle and its phenomena but one chapter—the object being to present the subject of Vital Magnetism in its practical, efficient and determinant aspect."

98 WILLIAM STREET, NEW YORK,
October 20th, 1877.

DEDICATION.

To my friend Dr. C. E. S., I respectfully dedicate the results of reading and observation embodied in these papers.

When you were the honored President of the State Medical Convention, and presented facts in your possession concerning the treatment of disease by Vital Magnetism, there was deep interest manifested in the subject; but the total absence of information concerning it, on the part of those who attempted to discuss it, you pointed out to me, and thus prompted me to supply the lack, from abundant materials which I had accumulated.

Hoping, my dear friend, that what is here presented may stimulate inquiry and open the treasures of this great storehouse of blessing, I am as ever,

Yours sincerely,

F. T. PARSON.

" There are more things in heaven and earth, Horatio,
Than are dreampt of in OUR philosophy."

"Whoso recognizes the unfathomable, all-pervading domain
of mystery, which is everywhere under our feet and among our
hands ; to whom the universe is an oracle and a temple, as
well as kitchen and cattle stall—he shall be called a mystic,
and delirious. To him, thou, with sniffing charity, wilt pro-
trusively proffer thy hand-lamp, and shriek, as one injured,
when he kicks his foot through it.

"Wert thou not born? Wilt thou not die? Explain me
all this—or do one of two things : retire into private places
with thy foolish cackle ; or, what were better, give it up ; or
weep not that the reign of wonder is done, and God's world
all disembellished and prcsaic, but that thou thyself art
hitherto a sand-blind pedant."

CARLYLE.

CONTENTS.

		PAGE
DEDICATION		5
PREFACE		9
Magnetism as a Therapeutical Agent		11
Status of the Practice in Great Britain—Views of Eminent Men		15
Status of the Practice on the Continent—History of Its Development		27
The Evidence of Reason and Common Sense		35
The Received Hypothesis—Dr. Ashburner's Theory of Polarities—Barth's Views : Mesmerism not Miracle		43
The Principal Therapeutical Effects—Evidence of Eminent Practitioners		57
Anæsthetic Effects of Vital Magnetism		76
On Qualifications of Magnetic Physicians		80
TYPICAL CASES :		
	The Case of Harriet Martineau	85
	The Extraordinary Case of Anne Vials	95
	A Double Phenomenon	101
	Case of Uterine Disease with General Nervous Derangement	103
	Case of Nervous Suffering Leaving the Patient Dumb	106
	Cure of Deafness of Long Standing	108
	Case of Spasmodic Asthma	109
	Case of Neuralgia or Painful Nervous Affection of the Heart of Many Years' Standing	110

PAGE

Case of Severe Nervous Disease . . . 112

Epilepsy 114

Epileptic Hysteria, with Lockjaw and Contraction of One Leg 116

Case of St. Vitus' Dance 118

Dr. Braid's Experience 120

Painless Amputation of the Thigh, &c. . . 124

English Practitioners and Advocates . . . 131

Prejudice, Ignorance and Charlatanry no Longer Permissible 139

Theorizing 143

The Testimony of Test 148

My Own Position 174

Sympathetic Influence . . . . . , . . 176

Dr. Newnham's Exposition of Principles and Plea for Investigation and Adoption by the Profession . 183

The Lessons of History—Shall we be Stupidly Timid or Bravely Wise?—A Warning to Obstructionists . 198

APPENDIX.—Dr. Elliotson's Adjuration to the Profession, and Dr. Esdaile's Remarkable Protest to the American Congress — Magnetism the First "Anæsthetic" 217

LITERATURE OF VITAL MAGNETISM 225

INDEX TO CASES IN THE "ZOIST". 231

PREFACE.

THE object of this work is to furnish, in a condensed form, the actual experience of experts, in the treatment of disease by Vital Magnetism, whose statements will not be disputed.

Purely mental phenomena in connection with this force are purposely omitted, as they lead to bewildering speculations, such as have swelled other works to great size, but leading only to confusion, disgust, or to the neglect of this important branch of science by medical men.

A class of enthusiasts calling themselves *healing mediums* have met with more or less success in the use of this agency, and have sought to wrest it into the support of a mis-named *spiritual philosophy.* They have often succeeded only in arraying the intelligent against a force which has been known from the earliest ages, and which has a wide and benign application to the ills of man.

It is monstrous injustice to permit these *pretenders* to usurp control of such an agency, and to frighten off men whose talents and acquirements fit them to investigate its facts and to arrange them into scientific order. For there is no doubt that healing effects are produced here as elsewhere, *by the most careful attention to conditions, laws and principles;* and such as fail to recognize these, can expect no more success in the use of Vital Magnetism, than those who neglect the laws that govern other imponderable forces can look for valuable results in their use. It is as much under the limitation of Natural Law as any other force, though its range is extensive.

It is exceedingly desirable that there should be a full and
impartial investigation of the merits and nature of this powerful agency, which has been hitherto discussed chiefly by men
who exalted some *phenomenal* feature into undue prominence.

Dr. Braid, of Manchester, England, a celebrated surgeon, a
painstaking investigator of magnetic phenomena, and a successful practitioner of the magnetic art, conceived that he had
discovered *fundamental* elements, when he had merely found
a new method of producing the primary magnetic condition.
He baptized his supposed original discovery with the name of
"Hypnotism," and Dr. Brown Sequard refers approvingly to
this term, and vainly strives to point out a distinction, that
does *not* exist, between this and Vital Magnetism.

While speaking thus, I recommend to all inquirers in this
department the very able work of Dr. Braid, entitled *Neuryp-
nology.*

My purpose will be gained if I draw attention to this
important subject from a natural and practical standpoint;
while the authors whom I mention will satisfy all those who
have speculative tendencies, and wish to push their inquiries
into the deeper mysteries of the subject.

Many of the laws of this ancient and fascinating science of
Vital Magnetism are well understood. In order that this
knowledge may be more generally diffused, and that men of
learning may have in their hands *facts* to stimulate them to
arrive at profounder principles lying at its base, this treatise
is given to the public. After seventy-five years of opposition
to the results of experiments and observation, these facts
were found to be stubborn and unanswerable; and the last
twenty-five years have witnessed their complete victory over
prejudice and mere argument; so that, to-day, the first scientists and physiologists write and speak of them, not only with
respect, but also with warmest praise.

VITAL MAGNETISM.

I.

THE various names under which Magnetism has been known or employed during a century are mainly these: Mesmerism, Animal Magnetism, Induced Somnambulism, Psychic Force, Nervous Force, Mesmeric Sleep-waking, Electro-Biology, Perkinism, Electrical Psychology, Pathetism, Nervo-vital Force, Psycopathy, Braidism and Hypnotism. Other terms have been applied as some new phenomenal feature came to the surface; but the force remained fixed and unchanged in all this fluctuation of terminology, because it is one of God's immutable agencies. But, in the bitter struggle over names, and methods, and their relative value, the facts of HEALING have continued uniform, and humanity has gratefully acknowledged their benefit.

Dr. John Elliottsen, in the *Zoist*, an English quarterly journal, published from 1843 to 1856, uses this language:

"Mesmerism has always been true. Dimly known for thousands of years, in barbarous and semi-barbarous countries, known as to some of its high results to many of the great nations of antiquity, though the knowledge was confined to the chosen few, it is only beginning to be seen in its various aspects and ramifications, and to assume the character of a science—a science of the deepest interest and importance—inasmuch as the phenomena of life transcend those of all inanimate matter, and the faculties of the brain—the mind—are the highest objects in the universe that man can study; and inasmuch as its power over the faculties of the body at large, and especially over the whole brain and nervous system, is immense, and therefore capable of application to prevent and remove suffering, and to cure disease far beyond the means hitherto possessed by the art of medicine."

If we trace magnetism along the ages its footprints are discernible in ancient literature, on the sculptured monuments of Egypt, in the traditions of Chaldea, India, Persia, Greece, Judea and Rome, upon the early records of Christianity, during the Middle Ages, and upon the clearer pages of modern history. In barbarous and half-civilized communities, men characterized its cures as *sacred mysteries*, as secret *remedies*, as something magical and occult. The unbroken stream of results in healing has flowed down, in obedience to the law that is equally perfect with the laws of light and heat, of gravitation and electricity. The facts are uniform and indisputable.

The experience of antiquity has been amply supplemented by the record of the last century, showing an unbroken advance, slow but sure.

The efforts of the schoolmen in the past to establish medical systems upon series of facts, involved as bitter a warfare upon each other as they have ever expended upon the practice of magnetism. At first they denied the facts, and when these could no longer be disputed, they denounced the practice as " magical," and as leagued with the *black art!* But, as the medical schools themselves solidifred into form, and intelligence spread, the fear of " magic " died out ; the cures by magnetism increased and were acknowledged, until, later on, Royal Commissions were appointed to investigate. These learned bodies at first found it difficult to be honest and unprejudiced, and reported adversely upon *side issues*, but finally with favor, in spite of the ravings of members grounded in old beliefs.

Attacks have now ceased, and to-day the first living physiologists commend the practice in highest terms. The main objection now urged, is the difficulty of finding properly qualified men to pursue the practice, as no amount of technical knowledge will make a successful magnetic practitioner in the absence of natural endowment. Time will remedy all this, and a sufficient force of operators will be forthcoming to cope intelligently with a vast number of so-called "incurable forms of disease," especially of a nervous character, to which magnetism is wonderfully adapted.

It is not claimed for this practice that it is a UNI-
VERSAL remedy: no one but a charlatan would put
forth such a claim; but, that it possesses great remedial powers, when properly applied, and when
the conditions necessary to success are fully met,
is unquestionable; and a great number of maladies
that cannot be relieved by medication will find
their solution here.

But the difficulty is, as a well-known modern
writer has said, in speaking of magnetic practice,
that—

" There is certainly something very revolting to
the pride of the present generation, to admit that
magnetism, the simple medicine of nature, can do
what science and art cannot effect; to give up many
favorite theories and plans of treatment—to lay
aside the haughtiness of established doctrines—
and to yield them before the simplicity of a curative action, which many have the power to impart,
though each has not the skill to direct;—and to revive the practice of thousands of by-gone years—
years which are only dimly seen through the long
vista of pity for the ignorance and barbarism with
which they are marked;—and then to allow the
practical value of these long-forgotten processes as
superior to the present results of science, by adopt-
ing them. These are difficulties that require no
small degree of moral courage to surmount."

This sentiment was uttered thirty years ago.
The world moves; the battle has been fought, the
victory won. The period from 1840 to 1852 marked

the grand struggle, and the support of such names as lend their splendor to truth, especially among medical men, effectively annihilated opposition; no public attack has since been made; pride of opinion has succumbed; the practice in Europe has become well settled; it is rapidly gaining respect here, where it is largely and successfully pursued as a specialty.

II.

STATUS OF THE PRACTICE IN GREAT BRITAIN—
VIEWS OF EMINENT MEN.

It must be said that, in this country, we are at least twenty-five years behind Europe in the application of Vital Magnetism in the treatment of disease. India established a Magnetic Hospital at Bengal, under the superintendence of Dr. Esdaile, in 1840, which was sustained by the subscriptions of the first British residents of that Presidency, in which hundreds of the most difficult and dangerous surgical operations were performed without pain under the anaesthetic effects of magnetism. Two hundred and sixty cases were there reported, many of which were amputations of a dangerous and painful character.*

A Magnetic Hospital was established in London in the year 1846, and for eleven years the most astonishing results were obtained from this prac- tice exclusively, and were duly reported through the columns of *The Zoist*, the journal devoted to that interest, and already alluded to as being edited by Dr. Baird.

The first officers of this institution were—

The Right Hon. the Earl of Ducie, *President.*

Vice-Presidents: Baron De Goldsmid, the Earl of Carlisle, R. Monckton Milnes, M. P., J. H. Lang- ston, M. P., and Rev. G. Sandby, Jr.

Treasurer: Mr. Briggs.

Committee: Dr. Ashburner, Dr. Elliottsen, Dr. Chandler, Dr. Flintoff, Dr. Clark, Dr. J. Hands, Dr. D. Hands, Dr. F. G. Johnston, Dr. Symes, Rev. Dr. Dixon, Major Buckley, Captain John James, Mr. Blythe, Mr. Fradelle, Mr. Kingdom, and Mr. Top- ham.

Its operations for these eleven years were sus- tained entirely by popular subscriptions, and among its patrons were many of the highest dignitaries of Church and State.

The Earl of Ducie was a nobleman of high posi- tion, at the time being Lord-in-Waiting to Her Ma-

---

* It will well repay perusal if my readers can obtain a copy of Dr. Esdaile's *Mesmerism in India*, published by H. Baillière, London, or a copy of the reprint issued in this country, I be- lieve at Hartford.

jesty, and a member of Dr. Cumming's Scotch Church.

Five years later, after a prosperous career of the institution, I find The Most Rev. the Archbishop of Dublin President, with the Earl of Stanhope and the Earl of Dunraven added to the Board of Directors.

For the grand results obtained I must refer to their published annual reports, which would, if reproduced, challenge universal attention and respect.

It will be seen from this that men of the first reputation in the realm stood sponsors for the practical character of this noble science.

At the fifth annual meeting, they say, in reviewing the stormy time through which they had passed :

"The world is becoming every day better informed as to the reality, the curative and assuaging power, and the innocence of Mesmerism; and the English medical profession is now emancipating itself from the unfortunate feelings which overpowered it for the time twelve years ago."

A similar institution was established at Exeter, and in one of their reports Dr. J. B. Parker, resident surgeon, says: "I have performed over two hundred surgical operations without the patient's feeling the pain whilst under the influence of Mesmerism, including twenty most painful operations on the eye, tying the radial artery, more than one hundred bleedings, cutting off a very painful wart, and the extraction of upwards of forty teeth." The name and residence of each of these patients is given in full, with particulars of each case.

In 1850, another institution of a similar character was opened at Bristol under the name of " The Bristol Mesmeric Institute," and under similar auspices. Dr. Stevens, in the second annual report to the Society, from which I can only make a very brief extract, says:

" When it is considered that almost all the cases received have been of the most extreme character —some of years standing—and that many of them have been rejected as hopeless under ordinary treatment, the actual amount of benefit conferred ought to be the only just criterion. Amongst these cases are comprised several of the most severe forms of epilepsy, of paralysis, of hysteria, of extreme nervous debility with depression ; cases of tic-douloureux, with rheumatism, in all its varied shapes; spinal disease, with scrofulous affections of the hip and knee joint.

" There have also been several cases of nervous dyspepsia, sleeplessness, nervous headaches, acute inflammation of local parts, such as the eye, ear and throat, which have been subdued by mesmeric influence. Several minor painless operations have been performed, such as the extraction of teeth, and the application of caustic to various parts of the body."

Dr. Lane, in seconding the adoption of the report, said :

" It was not a little consolatory to find that Mesmerism was wending its way to the shores of the Bosphorus, and returning to the country of the Ptolemies, in which its great discovery was first developed—Egypt—2,500 years ago. The paintings in the tomb or pyramid of Sesostris speak in higher or lower strains of its adoption and practice in the curative art, by the ancient Egyptians. Mesmer only revived it.

"The Egyptians had an overwhelming example, in the case of Miss Martineau, the celebrated and useful writer, whose sedentary habits produced such derangement of the system as no medicine could reach; her physician left her, intimating that nothing remained but to smooth the descent to death. Here, then, were genius, and taste, and talent consigned to the tomb, when Mesmerism came, and with a bland and reassuring smile tendered its aid. In a month, her nightly doses of narcotics were rendered nugatory; excruciating pains subsided; the animal functions gradually returned to their wonted vigor; and, in about a twelvemonths after, she was discovered writing a letter on the Great Pyramid, having ascended to its summit without anything more than the ordinary fatigue of climbing to so vast a height, over steps at least two feet and a-half high! What a mighty triumph for Mesmerism !

"But, still greater remained; to get down is infinitely more difficult than getting up—a task fraught with pain and much danger. *Men* glide from step to step by the seat, but women prone on the chest;

and then they have to jump from six to eight inches
to reach the step below, and woe to the she or he
who shall lose their equilibrium ; bang, bang, you
go from step to step, till you reach the bottom a
shapeless and crushed mass ; and all this was accomplished by Miss Martineau by Mesmerism ! "*

Dr. Barnes, surgeon of Bath, who was present
at the meeting, said he had " been in the medical
profession twenty-six years, and for a long time
had had strong doubts of the truth of Mesmerism.
His first impression was received from a visit to a
woman in a state of madness, whom two powerful
men were unable to hold. He advanced to her,
fixed his eye upon her, took her hand, and she at
once fell down under his influence.

"Afterwards he was sent for to a man whom he
found with one side completely paralyzed, his arm
and leg being perfectly lifeless. He made some
passes over his person, the capillary vessels began
at once to fill, the warmth returned, and he was
soon restored, and a few days ago carried two hundred weight for a mile and upwards. Four days
ago he (Dr. Barnes) was called to see a child who
had been in fits for a fortnight, and under the care
of the hospital surgeons. She was insensible, kept
incessantly rolling about her head in a shocking
way, and had not spoken for many days. The moment that he touched the top of her head the roll-

---

* See Section IX., " Typical Cases," for the interesting details
of this memorable case, as given in Miss Martineau's own ver-
sion.

ing motion was stayed; she soon became perfectly still, and smiled at him, and directly afterwards she spoke. He left her greatly recovered, and the next morning her mother told him that in an hour after his leaving her she asked to be dressed, and was as perfectly well as ever she was in her life. This might seem to be a miracle, but it was the result of a principle which could be explained upon scientific grounds. He had produced similar effects in fever, and in lumbago and sciatica."

Another Mesmeric Institute was established in Dublin, in the year 1852, under the patronage of His Grace the Archbishop of Dublin, of which Falconer Mills was president. The following prospectus clearly sets forth its objects and the principles adopted for its guidance. It is typical of all the institutions named in this connection in utterly repudiating the superstitious assumptions of supernaturalism, and placing the practice upon a purely scientific basis:

"The Committee of the Dublin Mesmeric Association, in appealing to their fellow-citizens in behalf of a Mesmeric Hospital, feel themselves called on to explain the character and objects of the Association, and to remove some misconceptions which prevail, or might be likely to arise, on the subject. They wish it to be distinctly understood, that their object is not to maintain or to devise any scientific THEORY, but simply to extend the knowledge of certain important FACTS, and to facilitate the beneficial application of that knowledge, in the relief of suffering, and the cure of disease. With this view, they propose, not to occupy themselves with a variety of experiments for the gratification of mere SPECULATIVE CURIOSITY, but to seek exclusively for PRACTICALLY USEFUL results.

"Without intending to disparage scientific investigation, or to pronounce for, or against, any particular theory, they propose to confine their attention to what is practical and practically useful.

"But, they would, at the same time, remind those who are eager after philosophical speculations, that in all departments of knowledge the ascertainment of facts has always come first in order, and that correct theories, to account for existing phenomena, have always (when such HAVE been attained) appeared long after, and have been based upon those facts. The reversed procedure—that of seeking in the first instance for a theory, and disregarding, or misrepresenting all facts, that cannot be satisfactorily EXPLAINED on such theory—this is precisely that 'ANTICIPATION OF NATURE' which Bacon protested against as the bane of true philosophy, and which it was the great object of his life to supersede, by the 'INTERROGATION OF NATURE.'

"The Committee beg also to be understood, most clearly and candidly, that they do not seek to produce Mesmerism before the public as an agent capable of curing all diseases, or by any means wish it to usurp the place of the medical man; but simply put it forward as an individual remedy, capable of controlling and exercising a powerful and beneficial influence over a vast number of ailments; and more especially those called nervous and chronic affections, which medical men have at all times acknowledged as little benefitted by the mere administration of drugs.

"The Committee feel it of vital importance to their own character, to the people, and to the cause of Mesmerism itself, to disclaim any connection or co-operation with those who receive or practice Mesmerism, as derived from *supernatural* agency, or miracle-working power, independent of the laws of creation; but, on the contrary, would more humbly and gratefully acknowledge it as an additional means granted by the Almighty to man, to make him more dependent on Him, and more useful to his fellow-man ; and believe its phenomena to be solely the result of the vital force, conveyed by the operator to the more delicate and susceptible nervous organization of the person mesmerized ; just as light, or heat, or sound, &c., variously affect the animal or organic tissues they come in contact with.

"Lastly, the Committee would suggest to the public, and in particular to the medical men of this city, that, when the existence of a real and important agent has been established by facts which it is vain to attempt denying, and when great and increasing public attention has been drawn to these facts, it is most important to the well-being of society that such an agent should *not be left in the hands of the ignorant and thoughtless, or of designing or ill-disposed persons;* but that means should be taken for affording all candid persons an opportunity of fair investigation ; so that truth may be distinguished from falsehood, and a beneficial from a noxious or dangerous application of the powers which Providence has placed within our reach.

"By Order.

"J. MacDonnell,

"*Hon. Secretary.*"

The grand results obtained in this hospital would be exceedingly interesting as related in their published reports, but space will not allow even a synopsis of them here. The same is true of the reports published at the various annual meetings of each of the other institutions named. They teem with interesting and startling records of cases ; the meetings themselves were always of marked interest, and in numbers full to overflowing, hundreds at times being unable to gain admittance.

Statements enough, I trust, are here submitted to show the strong position this science has assumed in the British Islands ; and in these entire reports we find neither rhapsody, cant, nor overstatement, but a judicial calmness, weighing carefully the aggregation of facts by men thoroughly competent to pass a solid judgment upon them.

Dr. Elliottson, in presenting the sixth annual report at a meeting presided over by the Earl of Dunraven, said :

"It may cure as well as any other remedy, and, what is very important, without pain or discomfort; it may cure when every other remedy fails; it may, without pain or discomfort, lessen disease which it does not cure; and it may prevent and lessen suffering when it does not lessen disease.

"When we possess established means of curing or alleviating disease or lessening distress, and above all, in cases over which they exert their beneficial operation without suffering or mischief, and in cases in which they cure better than Mesmerism, they ought to be employed as the profession at large employs them, and as we, who are medical men do, in common with our brethren, employ them in our private practice; though we give not a particle of any drug in the Infirmary, for patients enter on the understanding that they are to be treated with Mesmerism only, and we take no cases in which medicines are likely to be required ; our object is to supply to the poor that which is denied them in other hospitals and dispensaries, and, if we employed anything but Mesmerism, our cures would be ascribed by the profession to the medicines and not in any degree to the Mesmerism.

"In ordinary practice, if at the same time Mesmerism ALSO is employed, all established means will tell the more ; they will require to be used with less force, the cure will be speedier and pleasanter, and the recovery both from the disease and from the debility left by it, and very many ordinary measures, be much accelerated.

"The advantage of Mesmerism during convalescence from any disease and after any treatment, is
also very great. It often serves all the purpose of
change of air and scene, and is a blessed substitute
to those whose means are too limited for that
advantage. Wonderful as the fact may appear,
persons very susceptible of mesmeric influence, if
from confinement or other debilitating causes they
become languid, pale and uncomfortable, lose their
appetite, and are altogether in a state for which
a change of air and relaxation from labor are an
almost certain remedy, find mesmerization a far
more potent means."

Is any better authority needed than this great
physiologist, and one of the first physicians in the
city of London?

See how carefully he guards all the interests of
the profession he adorned, while stating such substantial facts as came under his observation during
the many years he occupied in the investigation of
Mesmerism.

Equally significant is the statement of that eminent representative of the Homœopathic School,
Dr. J. J. Garth Wilkinson, of London, in which he
declares: "We have no certain knowledge of the
limit of Mesmerism as a curative agent, nor of the
conditions which should EXCLUDE cases from this
treatment." This is certainly a very near approach
to Dr. Mesmer's universal remedy.

And a long stride is assuredly made in the recognized standing of magnetic practice, when that

greatest of living phyisologists, Dr. Wm. B. Carpenter, of England, can say, without provoking comment or contradiction, that this force, when applied with skill " *will take rank as one of the most potent methods of treatment which the physician has at his command.*"*

III.

STATUS OF THE PRACTICE ON THE CONTINENT—

HISTORY OF ITS DEVELOPMENT.

IN France the practice has been recognized by the great majority of its physicians, and, previous to the discovery of ether and chloroform, was ex-

---

* It is greatly to be regretted, however, that this valuable testimony of Dr. Carpenter's should be to some extent neutralized by his recent attempt in a popular magazine to excite ridicule over the mistakes of Dr. Mesmer, passing by the immense good he accomplished in the very line commended above, during his long and useful life.

What writer or discoverer in any department of science is exempt from the follies of immature judgment? Is it *fair*, then, to represent Mesmer as holding to opinions and practices which he revised and abandoned long years before his death?

Why draw imaginery lines, too, in regard to Mesmerism, and Hypnotism, where they do not exist, if the desire is sincere to see this practice extended?

tensively employed as an anæsthetic in capital cases of surgery in Paris hospitals.

As early as 1834 an organized Magnetic Hospital was opened by the Marquis de Guibert, a French nobleman,upon his estate at Fontchâteau in the Commune of Tarascon, and during the six years which this benevolent gentleman gave to the work, 3,315 cases were treated for severe forms of disease which had mainly been pronounced incurable. Dr. Charpignon, member of the Medical Faculty of Paris, in his work entitled " Phisiologie Medicine *et Mataphysique* du Magnetisme," in speaking of the results obtained by the Marquis in this hospital, says:

"Out of 3,315 patients admitted, 1,948 were discharged cured, and 375 were partially relieved, and 504 were impassive. Of those admitted 2,121 were females, and 1,194 were males; amongst other cases treated 14 were for contracted limbs, of which seven were perfectly cured, and others relieved "—(p. 274 of the above named work).

I pass by the thousands of cases treated by Dr. Mesmer, and by his coadjutor, Dr. D'Eslon, as they are matters of permanent history, and challenged at the time the attention of all Christendom, provoking the appointment of Royal and Scientific commissions for the investigation of their merits, and an offer of a large sum by the French Government to Mesmer to disclose the secret of his wonderful power over disease, of which he claimed to be the sole discoverer.

The world has been largely benefitted by Dr. Mesmer's researches in magnetism; and humanity will remain his debtor therefor to the end of time, and can well afford to forgive the folly of such exclusive claims as he set up respecting the discovery itself. The science also was rescued by his efforts from the domain of the so-called supernatural, and placed upon something like a scientific basis. Like all students and experimenters treading in new and unknown paths, Mesmer made mistakes; he surrounded the practice with a vast paraphernalia which did not belong to it, and made claims which could not be adequately supported; but these were of minor importance and were abandoned by this grand student when he became convinced of their inutility; (and that is the strongest expression which can be fairly used in regard to them). The MAGNETIZED RODS and the BAQUET were abandoned and the more natural methods of his coadjutor, Dr. D'Eslon, took their place.

These changes, however, did not affect the main question of CURE ; suffering humanity was relieved of pain and distress, while the SAVANS quarreled over methods.

The opposition to magnetism at this era, while bitter in many quarters, only passed through the same stages which it assumed toward Harvey on Circulation, Jenner on Vaccination, and the discovery and use of Peruvian Bark ; and like those discoveries, magnetism just as completely gained its victory ; until to-day the apparently arrogant claim of Mesmer, that " *There is but one health, one disease, and one remedy,*" has far more of truth than fiction in it.

Colquhoun, in referring to the antiquity of vital magnetism, says :

"In 1666, or more than a century prior to Mesmer, Valentine Greatrakes, an Irish gentleman, appeared in England, who posessed a power and wielded an influence quite equal to Mesmer. In the published accounts, Dr. George Rust, Lord Bishop of Derry, Ireland, says of the cures performed by him: 'I was three weeks together with him at my Lord Conway's, and saw him, I think, lay his hands upon a thousand persons ; and really there is something in it more than ordinary ; but I am convinced it is not miraculous. I have seen pains strangely fly before his hands, till he hath chased them out of the body—dimness of the eye cleared, and deafness cured by his touch ; twenty persons, at several times, in fits of the falling sickness, were in two or three minutes brought to themselves, so as to tell where the pain was ; and then he hath pursued it, till he hath drawn it out of the extreme part ; running sores of the king's evil dried up, and kernels brought to a suppuration by his hand ; grievous sores of many months date healed in a few days. Obstructions and stoppages removed, cancerous knots in the breast dissolved, &c.' 'But yet,' continues the Bishop, 'I have many reasons to persuade me that nothing of all this is miraculous ;' and he then proceeds to state his reasons. Joseph Glanville, the author, also published an endorsement of Greatrakes, and gave many interesting facts in detail of his cures, and also quoted several 'sagacious and wary persons of the Royal Society, and other learned and judicious men whom we may suppose as unlikely to be deceived by a contrived imposture as any others whatsoever.' "

It provokes a smile at this late day to see this balancing of probabilities of MIRACULOUS cure ; but these footprints in the past are interesting and convincing, nevertheless.

In one of the volumes of the German Archives of Animal Magnetism (Archiv für Thierischen Magnitismus), an account is given of John Joseph Gassner, born in 1727, at Bratz, in Suabia, who created a great sensation by his magnetic cures, especially of spasmodic and epileptical complaints. In the same work (Vol. I., No. 3) is published an account of a Magnetic Physician by the name of Richter, in 1817, whose cures were considered very wonderful, and were effected by magnetic manipulation. He was visited, it is said, by multitudes, " from ministers of state and noblemen down to the lowest beggars, and he cured them all, indiscriminately and gratuitously." The Government had his conduct investigated, and granted him protection.

In 1492, Petrus Pomponatius, Professor of Philosophy at Padua, assumes it as a fact generally acknowledged, that there are men endowed with the faculty of curing certain diseases by means of an effluence or emanation which they direct towards the patient.   He says " when those who are endowed with this faculty, operate by employing the force of the imagination and the will, this force affects their blood and their spirits, which produce the intended effects, by means of an evaporation thrown outwards," and that it is by no means inconceivable that health may be communicated to a sick person by this force of the imagination, and the will so directed.   He compares this susceptibility of health to the opposite susceptibility of the infection of disease.

These views and practices, be it remembered, were put forth nearly 400 years ago,—the PRACTICE to-day remains identical,—the THEORIES are not much improved.

Van Helmont has occupied so prominent a place in historical records, that a brief reference to him will be acceptable.   John Baptist Van Helmont was an eminent physician, who lived between 1577 and 1644.   He discovered the laudanum of Paracelsus, the spirit of hartshorn, and the volatile salts ; and to him we owe the first knowledge of the elastic aëriform fluids, to which he gave the name of GAS, which they still retain.

Van Helmont wrote a remarkable treatise on the "*Magnetic Cure of Wounds,*" in answer to two authors, who have written upon the same subject : Goclenius, a physical philosopher in high repute, and Father Robert, a Jesuit.

The first had maintained the reality of the cures effected by magnetic means, ascribing them to natural causes. The latter admits the cures, but ascribes them to the influence of the devil. Van Helmont wrote in order to put the practice upon a natural and scientific basis, deeming the defence of Goclenius feeble, and to prove, in opposition to Father Robert, that there was nothing diabolical in magnetic treatment. Magnetism, he says, is a uni-.versal agent ; there is nothing new in it but the name, and it is a paradox only to those who are disposed to ridicule everything, and who ascribe to the influence of Satan all those phenomena which they cannot explain. " Magnetism is that occult in- fluence which bodies exert over each other at a distance, whether by attraction or impulsion." The medium he designates " MAGNALE MAGNUM," and describes it as pervading all nature. Our modern theorists have not advanced much beyond this writer of 300 years ago.

One of the grand practical uses this great and good man made of the magnetic cure, was its successful application to the sick with the plague which raged in Brussells. " Perceiving," he says, "that most of the physicians deserted the sick, I devoted myself to their service, and God preserved me from the contagion. All, when they saw me, seemed to be refreshed with hope and joy."

This is part of the record of three hundred years ago.

Perhaps I have given references enough to establish the high antiquity of magnetic practice ; but if the reader would pursue this branch of the subject further, he will find himself abundantly gratified by a perusal of Colquhoun's *Isis Revelata*, published in Edinburgh in 1844, in which will be found an almost connected history, running back to the first century,—to which interesting work I am indebted for many references. That work, however, is somewhat controversial in character, and answers objections as to the REALITY of magnetic phenomena. That necessity having passed away, the book has gone out of print ; but it is one of the very best on the subject in our language.

/ It will be seen that this practice ante-dates that of medicine by many centuries, and has received the endorsement of eminent physicians in other ages, for it in no wise antagonizes, but *harmonizes* with the intelligent practice of medicine, and reaches a great variety of diseases for which no other successful treatment is found.

IV.

One of the objects in presenting the facts set forth in this treatise is to call the attention of physicians in this country more definitely to a subject which has, within the past few years, occupied the attention of some of the Medical Conventions of several of our States, only to exhibit the almost total lack of information at the command of the profession, while strong curiosity has always been manifested by the profession in regard to these processes of nature whenever the matter has been brought forward. The practical results to be attained by magnetic treatment in epilepsy, hysteria, paralysis, fever and ague, and the whole range of neural disorders and functional derangement, are scarcely dreamed of by the ordinary practitioner.

This ignorance does not arise from a low standard of learning generally; but physicians in this country have considered magnetic phenomena a curious *mental* condition without practical utility in general therapeutics, and have been content to hold vague and imperfect opinions. It is only when brought face to face with the researches of investigators abroad who have had the field almost to themselves, that they can understand the interest which men like Dr.Wm. B. Carpenter and Dr. Brown Sequard express in their public utterances upon this subject.

Physicians excuse themselves under the plea, as was said to me recently by a medical man, that "a man cannot be expected to know everything;" which is true enough, but, while the widest and deepest knowledge and most careful training would not be wasted, and would be as useful in this practice, as in surgery, a superficial acquaintance even with its facts and observation of its results will lead a practitioner possessing moderate ability in this direction, to relieve special patients from many a fearful experience which medication might be powerless to reach.

Outside aid might then be called in to complete the cure in a case that proved too exhausting for the general practitioner to pursue. All this would be in the same line as the employment of experts in SURGERY where the skill of the attending physician is often unequal to the safe performance of some delicate operation, and properly gives way to more skillful hands.

One reason for the lack of progress in the field of magnetic therapeutics in this country, lies in the fact that the art has often been degraded into a mere spectacle by itinerant lecturers, who seized upon the novel mental phenomena, so easily excited in a certain percentage of natural sensitives, and presented the buffoonery elicited from these 'hypnotized' automatons as proper illustrations of the real power of animal magnetism. The natural result was to discourage and prevent investigation into its real merits.

·But, as we have recorded, this has not been the fate of this noble science abroad, at least not to the same extent. It had to struggle for a time, as we have seen, against hate, opposition, calumny and misrepresentation ; even the first report of the French Academy in 1784 was against it; still it advanced steadily, until the second investigation ordered by the French Royal Academy of Medicine, in 1825, when opposition was to a great extent silenced.

To this important commission let us now more fully advert. It was composed of the following eminent physicians : MM. Bourdois, Double, Fouquier, Itard, Guéneau, de Mussy, Guersent, Leroux, Majendie, Marc, Thillaye and M. Husson. Six patient years of investigation were expended upon this subject, and a wide range of experiments instituted, many of which were perhaps of little practical value, covering CLAIRVOYANCE, INTUITION, and INTERIOR PREVISION, culminating in a report, too long to be embodied here, of these interesting phenomena, but incorporating this practical conclusion, that, *"Considered as a physiological phenomenon, or as a therapeutic agent, magnetism should be assigned a place in the list of medical knowledge."*

The report throughout is a candid, able, and clear statement of fact, and, after detailing some cases of cure, closes as follows: —"We do not ask of you a blind assent to all which we have reported. We can conceive that a great number of these facts are so extraordinary, that you cannot yield to us your belief;—perhaps indeed we ourselves should have dared to refuse you our assent, if, having changed places, you should have announced to us this day, from this chair, these very facts which we had never seen, never observed, never studied, never pursued.

"We ask only that you would judge us as we would judge you ;—that is to say—that you would feel thoroughly convinced that neither the love of the marvelous, nor the desire of celebrity, nor any personal interest, has influenced our labors. We have been animated by higher and worthier motives ; by the love of science and by the desire to justify the hopes which the Academy had formed of our zeal and our devotedness."

This report was signed unanimously, and redeemed the Academy from the poltroonery which the report of 1784 carried upon its face.

During the last century the best writers upon the subject have come from the ranks of the medical profession, embracing men of high standing ; clergymen of repute, and thorough scientists have also contributed to its literature ; and yet how little is known in these United States of the capabilities of this grand force ! Where are our writers ? Scarcely one worthy of the name has appeared : while hundreds of volumes of valuable works upon this subject have been poured forth in France, Germany and England.

In 1854-'5 more books were published in France on the subject of vital magnetism than upon any other topic ; many of which, it is true, were of a highly speculative character, undertaking to explain the inexplicable, but many were of rare value.

It will not be my purpose to attempt to explain the origin or causation of this force ; the difficulty of such a task is too apparent, and is well set forth in the language of Dr. Roget's Bridgewater Treatise, when he says :

" In the investigation of the powers which are concerned in the phenomena of *living beings*, we meet with difficulties incomparably greater than those that attend the discovery of the physical forces by which the parts of inanimate nature are actuated.

" The elements of the *inorganic* world are few and simple ; the combinations they present are, in most cases, easily unravelled ; and the powers which actuate their motions, or affect their union and their changes, are reducible to a small number of general laws, of which the results may, for the most part, be anticipated and exactly determined by calculation. What law, for instance, can be more simple than that of gravitation, to which all material bodies, whatever be their size, figure, or other properties, and whatever be their relative positions, are equally subjected ; and of which the observations of modern astronomers have rendered it probable that the influence extends to the remotest regions of space? The most undeviating regularity is exhibited in the motions of those stupendous

planetary masses, which continually roll onwards
in the orbits prescribed by this all-prevading force.
Even the slightest perturbations occasioned by their
mutual influence, are but the direct results of the
same general law, and are necessarily restrained
within certain limits, which they never can exceed,
and by which the permanence of the system is
effectually secured. All the terrestrial changes
dependent on these motions partake of the same
constancy. The same periodic order governs the
succession of day and night, the rise and fall of the
tides, and the return of the seasons ; which order,
as far as we can perceive, is incapable of being disturbed by any existing cause.

"Equally definite are the operations of the forces
of cohesion, of elasticity, or of whatever other
mechanical powers of attraction or repulsion there
may be, which actuate at insensible distances the
particles of matter."

After observing that all these phenomena, together with those of chemistry, light, heat, electricity and magnetism, have been, in like manner,
reduced to laws of sufficient simplicity to admit of
the application of mathematical reasoning, and
that to whatever department of physical science
our researches have extended, we everywhere meet
with the same regularity in the phenomena, the
same simplicity of the laws, and the same uniform-
ity of results, the author continues :

"Far different is the aspect of living nature. The
spectacle here offered to our view is every where

characterized by boundless variety, by inscrutable complexity, by perpetual mutation. Our attention is solicited to a vast multiplicity of objects, curious and intricate in their mechanism, exhibiting peculiar movements, actuated with new and unknown powers, and gifted with high and refined endowments. In place of the simple combinations of elements and the simple properties of mineral bodies, all organic structures, even the most minute, present exceedingly complicated arrangements, and a prolonged succession of phenomena, so varied and so anomalous, as to be utterly irreducible to the known laws which govern inanimate matter.

" If we are to reason at all, we can reason only upon the principle that for every effect there must be a corresponding cause ; or in other words, that there is an established and invariable order of sequence among the changes which take place in the universe. But though it be granted that all the phenomena we behold are the effects of certain causes, it might still be alleged, as a bar to all further reasoning, that these causes are not only utterly unknown to us, but that their discovery is wholly beyond the reach of our faculties.

"The argument is specious only because it is true in one particular sense, and that a very limited one. Those who urge it do not seem to be aware that its general application, in that very sense, would shake the foundation of every kind of knowledge, even that which we regard as built upon the most solid basis.

" Of causation, it is agreed that we know nothing : all that we do know is that one event succeeds another with undeviating constancy."

This philosophy covers the demand for an EXPLA-NATION of the laws governing the phenomena of magnetism.

There would be no more justice, however, in rejecting the established effects of this force, than those of either of the forces here alluded to ; and realizing the utter futility of attempting to answer unreasonable demands, I pass by the sea of speculation launched upon by various writers, and confine myself to arraying some of the prominent facts in support of its claims bearing upon the welfare of the human race, and turn over to the common sense of the reader all objections in this direction, including the weaker ones, of diabolic agency, miraculous power, spirit mediumship, or any supernatural agency.

Let us accept the fact that the Almighty Father has placed at our disposal a sanative force that will assuage pain and cure disease by a natural process that is *always safe, never injurious,* and adapted to rescue many a weary and hopeless sufferer from the region of despair.

V.

As a practical investigator and practitioner I may
be expected to have balanced the theories of others,
and arrived at some philosophical conclusion bearing upon cause and effect.

In order to state this clearly, I quote a brief
extract from Dr. John Ashburner's "Animal Magnetism," as setting forth in the most condensed
form, what I deem to be a *reasonable* hypothesis, and
which my own experience and observation go far
to supplement and confirm.

And yet, simple as this theory appears to read,
it is by no means certain that all its points can be
absolutely *proved;* still, the facts and the theory are
quite as susceptible of demonstration and proof, as
those pertaining to Light, Heat, and Electricity, and
in the absence of anything more definite they may
be accepted. If the arguments by which he arrives
at his conclusions are not conclusive, they certainly
are convincing. He says:

"Man is a magnet. He has, like all other magnets, poles and equators. But, being a magnetic

machine of very complex structure, his magnetic apparatus is divided into many parts. The brain is the chief magnet, and the trunk and extremities are separate magnets, having intimate relations with the chief source of magnetism. We infer from these facts, what is the truth, that the normal currents take a normal course from the brain to the caudal extremities. Mr. Faraday and Signor Matteucci have already established this fact in the gymnotus and in the torpedo.

" Polarity belongs to magnets ; every part of the living body is a magnet, and every magnet is capable of being submitted to a reversal of polarity.

" Those who are conversant with the common instrument called the galvanic battery, may easily convince themselves of the fact of the reversals of polarity, by uniting the extremities of the positive and negative wires from the two poles of a battery properly charged with the fluid, calculated to set in motion the electric current. At the moment of contact the fluid in the cells effervesces. The polarities of the current have been reversed. This fact of the reversal of polarity, applied in the case of the living body, explains the phenomenon of pain.

" While the animated being proceeds in a direct train, his sensibilities are not the sources of conscious feeling. They are to all intents quiescent. The polarities of the magnet are undisturbed. In all directions the ultimate organic molecules, constituting the organism, are undisturbed. This is sleep. The moment the relations of these mole-

cules to each other are altered, the individual becomes conscious of the change by an altered state
of sensation. This is not necessarily a pain, or a
person could never become awake from a condition
of sleep without suffering pain. The conditions of
sleep and wakefulness are striking examples of altered opposite polarities. The whole body being a
congeries of magnetic molecules, must necessarily
be subject to the laws regulating polarities. Any
change in the relations of the poles of living animal
molecules, must be productive of a change in the
sensibilities of the part. Whether the change be
the cause of pleasure or of pain, must depend upon
the faculties of the individual.

" Endowed with a nervous system, the animal is
susceptible of sensations, without which, the idea
of pleasure or pain becomes absurd. The inference,
then, remains that *pain is an extreme disturbance of
the polarities of a part*," (pp. 88, 97).*

" To illustrate the application of this principle I
will give a typical case from the same author:

" I lately had a case of a young lady aged thirteen years, who had suffered for years from a suc-

---

*Should any of my readers desire to investigate exhaust-
ively the theory of magnetic polarities, they will find a thorough analysis of this physical phenomenon in Baron Von
Reichenbach's work, " *Researches on Magnetism*," translated from
the German and edited by Dr. Wm. Gregory, Professor of
Chemistry in the University of Edinburgh, and one of the best
writers on magnetic science. This work of Von Reichenbach's
contains about five hundred octavo pages, and is the result of
many years of careful experiment upon the human body, min-
erals, crystals, &c., demonstrating the presence and range of
magnetic and odyllic force over a very wide field in nature.

cession of strumous affections, of which the neck and left side of her face exhibited evidence. Marasmus with epigastric pain, aggravated by pressure upon the part ; rheumatic pains, first in one ankle then in the other, aggravated by the same description of pains in both wrists, and in several spots in the course of the spine, threatened to wear out the poor sufferer with want of sleep ; hectic fever with rapid pulse, distressing thirst and perpetual desire for relief, kept the attendant on the alert six or seven times in the night. The remedies usual in such cases had all been had recourse to in vain. Months rolled on and the disease marched without a prospect of alleviation. Additional advice was obtained, but no bright hopes could be indulged in.

" If the arthritic symptoms were those of strumous inflammation, why did not a serious disorganization take place ? There was abundance of pain in fibrous tissues, with slight increase of heat, and so much tenderness that the joints could not be handled. Month after month, pain, the extreme reversal of polarity, gave the local symptoms of premonition— the premonitory advanced guard of the enemy— inflammation ; the rest of the troop of symptoms had not yet joined their comrades, pain and slightly increased heat.

" That pain is dependent on a reversal of polarity, that the pain in analogous conditions of gout and scrofula is subdued by measures which restore the healthy polarities of organs, is obvious, from the fact that it is necessary only to mesmerize the pained parts for a sufficient time *downwards*, in order to overcome the suffering.

" The process of cure consists in establishing the healthy direct current in place of the inverse, or lethal, current.

" The young lady alluded to was submitted to downward passes for an hour at a time twice a day during three months, and she was cured ; restored by God's blessing of Mesmerism to such health as to be able to enjoy life with all the exhilarating influences and impulses of youth. She was a pattern of good health and spirits.*

In spite of the irritability of the doctor (for which he doubtless had good reasons) he makes a noble and convincing claim for the healing effects of Vital Magnetism, of which he was a worthy champion. In the preface of his extensive work he says: " We propose to show that Animal Magnetism is a force which can alleviate the most agonizing suffering ; which can cure painful cancers and other malignant diseases; which, without endangering life, can render the patient insensible to pain under the surgeon's knife, and in the throes of a painful labor; which can raise up from the deep insensibility of the last stage of typhus fever a dying patient; and which can do far more than all this, viz: cure by its vital and healing efficacy the frightful suffering

---

* That Dr. John Ashburner is competent authority none will question, and his work on magnetism is as comprehensive as any, especially in its application to disease. It is published by H. Baillière, 219 Regent Street, London, and was issued in 1867.

arising from an attack of gout in the peritoneal membrane of the abdomen, for which, I was told that fourteen of the most eminent physicians and surgeons had failed to suggest any means of relief. May we not assert, then, that through the agency of this force, God has enabled man to wield a power almost divine!" The author makes good these pledges.

Many years of patient study and experiment were expended upon his investigations by the eminent scientist and scholar, Von Reichenbach, who was most thoroughly furnished, in every way, for such a work.

I am aware that some men have flippantly declared that in all this painstaking research this grand man is self-deceived and deceiving : but, few of his critics have yet attained the high position occupied by Von Reichenbach, in Chemistry, Geology, Mineralogy and the domain of Physics: nor can they displace him from public estimation by their shallow doubting: his " Researches "—already adverted to—will live long after the memory of these jealous parasites shall have perished.

So thoroughly has he done his work that he is a bold man who will lightly undertake to traverse the same path.

Mr. G. Barth, an eminent magnetic practitioner in London, also a vigorous writer on magnetic science, advances these views, in his very interesting and suggestive little volume, " Mesmerism not Miracle," as the result of his own study and wide experience.

" Before we endeavor," he says, " to show how the

mesmeric agent cures disease, we may perhaps be allowed to speculate on what constitutes health and what disease, and what the mesmeric influence is.

"A man is in full health when every organ in his body is capable of duly performing its assigned use ; doing neither too much work nor too little work : but just fairly and equally sharing its own task in the general labor required from the commonwealth of organs, and doing this also at the proper moment, working when its proper turn comes ; each organ being sound and perfect in its structure and every vessel, fibre, tissue and atom, perfectly and properly constituted. Disease is just the reverse of this picture : we have organs either imperfect in structure and unable to perform their functions in consequence of such imperfection, or we have them, although not structurally altered, incapable or refusing to do their fair share of the general labor, or doing too much, and doing it at wrong times, and thus disturbing the general harmony of the commonweal. Structural disease, unless congenital or from some accident, rarely exists excepting as the consequence of a previous functional derangement. A question may arise here—What causes functional derangement?

" Functional derangement is a disturbance of the power, force, or agency which acts upon organs and makes them work. To understand the subject clearly we must endeavor to view the active agency or moving power as an entity, separate and distinct from the passive recipient or organism to be acted on. We shall not obtain much help by studying the subject in the method which authority ordains in medical schools. Mesmerism, or some analogous science, must be admitted and used as instruments of investigation before either physiologists or pathologists will be able to obtain a masterly view of a human being as a whole, or comprehend fully the laws on which depend his health and disease. Supposing that an anatomist-physiologist, a microscopist-physiologist, and a chemist-physiologist were determined by the aid of their general and particular knowledge to analyze a man thoroughly and make plain the whole matter. They catch a live man, weighing 150 pounds we may say, and as a preliminary step in their investigation they give him a dose of prussic acid, or take some other means of suspending the action of his heart; the anatomist then does his work, and shows bones, blood vessels, nerves, membranes, muscles, tendons, cartilages, fat, and a membrane investing the whole, namely, skin. He shows heart, lungs, stomach, liver, and various and numerous other organs in the body, the use or functions of each, and the healthy character of its structure he determines as a physiologist and pathologist. He shows a brain, spinal chord and system of nerves branching off in every direction, going from the brain as an imaginary centre to every one of these organs, to the extremities, and the whole skin as a circumference. The microscopist shows that these nerves, even the most minute, are tubes, and a rational in-

ference follows that being tubes they were intended
to contain and convey; his instrument discovers the
mechanical structure of the blood and other fluids
in the system and the radical molecules, or germs
from which all organs and tissues are developed.
The chemist takes up the analysis where these leave
it and reduces the whole to separate portions or
component elementary principles, called hydrogen,
oxygen, carbon, nitrogen, calcium, phosphorus,
and small parts of sundry other elementary
principles which need not be here specified, but
which altogether weigh 150 pounds;—the exact
weight of each, the proportions in which one com-
bines with another, and the character of every re-
sultant combination are next duly eliminated,
arranged, and tabulated.

"Weary and tedious—mentally exhausting and
sensationally disgusting—has the task of these deep
searching physiologists been; the reward of their
labors remains to them; they have analyzed a
man !—they now know all about him !—and their
knowledge will be set forth for the good of the
human race, for this, when combined with the
learned contributions of fellow-laborers, morbid
anatomists, pathologists, therapeutists, nosolo-
gists, pharmaceutists, and sundry other *ists*, must
furnish such a battery of medical science to
bear upon the subjects of human health and disease
—of the composition of a man and his re-com-
position—of the wearing out and the renewal of his
tissues and organs—of the nature of healthy struc-

ture and of unhealthy structure—that disease will no longer have a chance of establishing itself in the system. If it attempts to invade, instantly the physician's battery is set into action and the disease annihilated. Is this the fact? have the researches of our physiologists enabled medical sciolists to establish a perfect system of healing by which they can comprehend and absolutely master disease? Experience cries, No!—we are very little in advance of the physicians of two thousand years ago in this respect. Is our supposed scientific investigation then useless? Not so!—it may be worth the labor spent, it is well as far as it goes, but it is not all, it is not enough; our physiologists have been analyzing a *dead body*—not a *living man!*

" The nerves which the microscope show as tubes meant to contain and convey something, certainly lost the fluid which must have circulated through their ramifications when the anatomist stopped the action of the heart; for he saw it not when he divided a nerve—the microscope could not detect it; there was no trace of it discovered by the chemist; there was no evidence of its presence; it had therefore already escaped if it ever were there. Although the physiologist has not seen this power or fluid, he rationally infers its existence as derived from the brain, and he also knows that its operative energy or force is antecedent and necessary to the functional activity of an organ, for if the nerves which convey it from the living brain are divided that organ no longer performs its function.

"It signifies not by what name we call this force ; we are not sure that it is either electricity or magnetism—but nervous force or nervous fluid, the common names by which it is designated, are not misnomers, and we may, therefore, retain them when alluding to it.

"As the integrity of structure depends on function and the integrity of function depends on the nervous fluid, we may readily believe that the power which can control this nervous fluid in an individual is a power competent to regulate his disturbed functional action and also to modify structural derangement to a greater or lesser extent.

"It therefore follows that this power, properly directed, is an agent by which diseases may be cured.

"*To operate on the nervous force of another by our own nervous force is to exercise the mesmeric art.*

"The nervous force appertains to a living man ; it is not to be found existing in a dead body ; if a philosopher would understand either its origin, nature or laws, he must study it where it is, instead of seeking to observe it where it is not.

"This study leads us directly away from an investigation of the gross ponderable material elements of the body, although organized, to a consideration of the imponderable forces which organized it, and which, operating upon or through the nervous fluid, maintain and conserve it as an individual organic existence.

"This study is the science of mesmerism."

And then, after speculating on received theories, he adds :

"Having as briefly as possible endeavored to show how mesmerism cures disease, we may inquire if our readers *can* see anything approaching to miracle in a mesmeric cure ? Instead of being supernatural, cures by mesmerism are particularly and peculiarly natural, certainly as much or more so than cures effected by the dynamic power which is found existing in medicinal drugs.

"There is a natural law existing that all dissimilar electrical states tend to become similar states, if the atoms of matter which are in opposite conditions be sufficiently approximated. The plus or positive state, and the minus or negative state mutually attract until plus and minus cease to exist and equilibrium is established. We know not why this is—we only know that it is. The analogy will apply to two human organisms, one of whom is in a state of health, the other suffering from disease ; if they are brought into certain relations to each other, the one who is healthy or in the positive state will transmit his state to the one who is nega- tive or unhealthy until equilibrium is established.

"There is a natural law relating to the magnetic forces of steel magnets. If we take a weak or ex- hausted magnet and make passes over it in one direction with a strong magnet, either by drawing it in contact, or at a short distance, the strong magnet imparts some of its force to the weak one, and thereby renovates or strengthens it until equilibrium is established, or the weak magnet has received as strong a charge as its material ferruginous atoms are capable of retaining. Again, the analogy will obtain as relates to certain mesmeric operations—the healthy man is the strong magnet, the sick man the weak one; passes in one direction cause special effects—passes reversed or made in the opposite direction dissipate these effects. Man has also his polar relations—one hand and half being positive, the other hand and half being negative ; corresponding to the north and south poles of a horse-shoe magnet.

"We see a similar law relating to the imponderable fluid or force called heat. We do not know what heat is ; we do not know that it is until we have some experience of its effects ; but we do know that if two bodies possessing different increments of heat approach each other, the one which is plus gives off its heat to that which is minus, until equilibrium is obtained. Here again the analogy of the healthy man and sick man, and the return to equilibrium or mesmeric cure may be urged.

"We do not know what electricity is ; we do not know what magnetism is ; we do not know what heat is ; we do not know what the vital influence of a living body is. We judge alone of the existence of these influences, fluids, or forces by their effects ; truly ! under certain circumstances they become visible—and how visible ?—as luminous emanations, auræ, flames or light. The luminosity of electricity,

heat, and magnetism are not disputed by philosophers ; the luminous appearance of the life-power
or influence is as easily shown if we use the necessary instruments.

" The law of the tendency to states of equilibrium
being established as a natural law affecting the three
first-named agencies, why should the corresponding
law of a tendency to equilibrium under the necessary conditions in respect to the fourth or vital-force
agency be considered unnatural?

" We may assume that there is some kind of imponderable fluid generated in every living system,
resembling the electric or magnetic fluids, causing
heat—being perceptible as light, conductible by
the nerves, and existing, associated with every
living atom of the living body ; the material atoms
only having life and maintaining their organic state
during this association."

These observations are based largely on Reichenbach's discoveries; and where the results of his
researches are not admitted they may appear to
have little force, but to the practical magnetist the
conclusions are sound and the premises valid.
They are submitted here as a fair sample of the
direction inquiry had taken at that period in the
minds of writers who philosophized on physical and
mental phenomena in magnetic science. If these
suggestions do not settle the questions presented,
they at least leave them in no worse plight than the
pretentious writers of to-day leave the various
obscure phenomena of which they treat in their

mental physiologies, and which generally obtain popular credence. They suffer nothing by comparison.

VI.

THE PRINCIPAL THERAPEUTICAL EFFECTS—EVIDENCE OF EMINENT PRACTITIONERS.

Colquhoun describes the leading therapeutical effects as follows;

FIRST: A general excitement and strengthening of the vital functions, without any considerable *stimulus*, in the nervous, muscular, vascular, and digestive systems.

Persons who could not be strengthened by corroborant medicines of any kind, have been restored to health, from a state of great debility, in a short time, by means of the magnetic treatment.

The application of this remedy quickens the pulse, produces an increased degree of warmth, greater sensitive power, and mental cheerfulness.

The appetite and the digestion are increased; the bowels, which had previously been kept open by artificial means, now become regular; and the patient acquires a relish for such kinds of food as are good for him, and an aversion for such as are injurious. Animal magnetism also promotes all the other secretions. In those constitutional complaints which are peculiar to the female sex, it is the most certain, the most powerful, and most efficacious remedy hitherto discovered.

The treatment seems to operate principally upon the great concatenation of sympathetic nerves (*the plexus solaris*) and by means of their various ramifications, to communicate its influence to the rest of the system.

SECOND: It affords a gentle STIMULUS, pervading generally the whole surface of the body, by which all disturbed harmony and diseased local action are removed, and the equilibrium again restored.

In this way animal magnetism soothes the most violent action of the nervous system, the tumult of the muscles, and the over-exertion of the vital functions in the whole economy.

THIRD: It draws off the increased vital action from the diseased parts, and conducts it to others.

By this means a two-fold advantage is gained: in the *first* place, the excited action, or irritation, is carried away from the internal and more noble organs to those whose violent action is attended with less injury to the system; and in the *second* place, the salutary vital action is strengthened and increased in particular debilitated organs. The consequences of the magnetic treatment, therefore, are SOOTHING and STRENGTHENING.

In most instances, the agitation produced by the diseased organization is gradually allayed, until, at length, a perfect recovery is effected; because animal magnetism occasions a diminution and total removal of the existing cause of the morbid action of the nervous system.

Dr. Ashburner says: "Perhaps none of the diseases that have not wholly disorganized or destroyed important structures is able to resist the influence of repetitions of the powerfully toning force of mesmerism.

"It appears to re-establish the original order of Nature: and obliges parts which have not been too much altered in structure, to resume their healthy forms and relations. It normalises abnormalized organs. It replaces disease by health. This language .is very deliberately, very earnestly used, from a thorough conviction of its truth, and is not the least too strong, considering the facts on which I rely."

Chauncy Hare Townshend says of this force: "To serve either as a calmant or stimulant, according to the exigencies of the complaint which it is called in to combat, followed as it is in neither mode of its influence by heaviness or exhaustion, would alone give it the highest rank as a remedy, if we regard it as a remedy merely. In this point of view, how valuable appear its offices, how unmatched by those of any substance in the Materia Medica! Again:—the direct correspondence of mesmerism with the nervous system, gives it a marked superiority over all such grosser agents as must reach that delicate framework of life by a circuitous route. Of all remedies, this alone pours its benefits direct upon the very springs of sensation; and thus, in cases of deafness and blindness, which depend on nervous weakness (and I believe there are more such affections referable to this cause than is generally suspected), we possess a subtle means of acting efficiently upon that fountain-head of the calamity, to which neither drug nor couching-needle can find its way."

Dr. Esdaile, already quoted, in his work entitled " Mesmerism in India," after detailing hundreds of remarkable cases in his practice, in hospital, sums up his conclusions of the general effects, as follows :

"I conclude that mesmerism is a natural power of the human body.

" That it affects directly the nervous and muscular systems.

" That in the mesmeric trance the most severe and protracted surgical operations can be performed, without the patients being sensible of pain.

" That spasms and nervous pains often disappear before the mesmeric trance.

" That it gives us a complete command of the muscular system, and is therefore of great service in restoring contracted limbs.

" That its administration often acts as a useful stimulant in functional debility of the nerves.

" That as sleep, in the absense of all pain, is the best condition of the system for subduing inflammation, the trance will probably be found to be a powerful remedy in local inflammations."

Dr. Esdaile's experiments were directed mainly to painless surgery, although he reports the relief and cure of a large number of nervous diseases and inflammatory conditions.

No physiologist living stands higher than Dr. Wm. B. Carpenter, of England, whose writings upon physiology and kindred science command universal respect, and are adopted as the text-books in the first medical schools in the world. Dr. Carpenter's attention seems to have been directed more specifically to Mr. Braid's so-called Hypnotic Methods, and in his recent work entitled "Principles of Mental Physiology" (*Reprint by D. Appleton & Co., N. Y.*, 1875), page 609, says:

"There is nothing in the least degree incredible, therefore, in the phenomena which Mr. Braid recorded, many of which the writer himself witnessed.

"The pulsations of the heart and the respiratory movements may be accelerated or retarded; and various secretions may be altered both in quantity, and quality, of which the following is a striking example.

"*a*—A lady, who was leaving off nursing from defect of milk, the baby being thirteen months old, was hypnotized by Mr. Braid, and whilst she was in this state he made passes over her right breast to call her attention to it.

"In a few moments her gestures showed that the baby was sucking, and in two minutes the breast was distended with milk, at which, when subsequently awakened, she expressed the greatest surprise.

" The flow of milk from that side continued most abundant: and, in order to restore symmetry to her figure, Mr. Braid subsequently produced the same change in the other; after which she had a copious supply of milk for nine months.

" The removal of morbid deposits under the same influence, seems quite as well attested, and the Physiologist who holds with the illustrious Mueller, that 'an idea that a structural defect will certainly be removed by a certain act, increases the organic action in the part,' will see no inherent impossibility in the following statement:

" *b*—A female relative of Mr. Braid was the subject of a severe rheumatic fever, during the course of which the left eye became seriously implicated, so that when the inflammatory action had passed away, there was an opacity over more than one half the cornea, which not only prevented distinct vision, but occasioned an annoying disfigurement.

" Having placed herself under Mr. Braid's hypnotic treatment for the relief of violent pain in her arm and shoulder, she found, to the surprise alike of herself and Mr. B., that her sight began to improve very perceptibly.

" The operation was therefore continued daily, and in a short time the cornea became so transparent that close inspection was required to discover any remains of the opacity." (*Neurhypnology, p.* 175.)

And he further declares:

" The writer has known other cases, in which secretions that had been morbidly suspended, have bccn re-induced by this process; and is satisfied that, if applied with skill and discrimination, *it would take rank as one of the most potent methods of treatment which the physician has at his command."*

Dr. J. J. Garth Wilkinson, Member of the Royal College of Surgeons, England, and one of the first Homœopathic Physicians of London, says in his brilliant and advanced work, "The Human Body and its Connection with Man," pp. 368–369:

"Mesmerism emphatically gives new or other life to those who need it; and it does this by the mere form and attitude which the agent and patient assume relatively to each other.

"The human world is full of powers in a state of balance and indifference. Change the posture of anything therein and the whole has to readjust itself to a new balance; a rush of forces takes place and currents pass to and fro until the equilibrium is recovered. The moral and the physical are both under this statical law.

"We have no certain knowledge of the limits of Mesmerism as a curative agent, nor of the conditions which should EXCLUDE cases from this treatment. In functional disorders of the nervous system it is especially indicated, and as a number of diseases, even seemingly organic, spring from this root, it appears that it has a large field of application here. Hysteria, Epilepsy, Catalespy, and those other maladies in which the visceral motions predominate over the rhythmical or rational motions of the lungs, come very markedly under its benefits. But it is not in our power to lay down any rule for the distribution to it of cases generally; and therefore we wait upon experiment, which shows that the utilities to be derived from its employment are very extensive.

"We had almost forgotten to place to the credit of Mesmerism its introduction of a painless surgery, which is among the most brilliant discoveries of the age. The doctors were totally incredulous of this matter, until Ether and Chloroform came and did the same thing in a grosser shape.

"If there were shame in the world, they must have felt it, when they found how easy their impossibilities of a fortnight before had become. They doubted the testimony of honest men, where Mesmerism was concerned; they accepted the same facts when Chloroform produced them."

Not to cumulate authority and evidence, but rather to show the remarkably wide range of application, the celebrated Dr. John Elliottson gives this summary of cases of magnetic cure coming under his observation for a period of ten years:

1843.—Five cases of insanity.
Eight cases of St. Vitus's dance.
Six cases of palsy of sensation or motion.
Seven cures of epileptic and other fits.
Rapid cure of delirium from grief.
Functional affection of the heart of seven years standing.
Six cures of rheumatism.

A few of these cases occurred in University College Hospital.

1844.—A remarkable cure of hiccough.
A cure of *epilepsy* of fifteen years standing.
A cure of obstinate spinal irritation.
Cure of a diseased knee in a child.
Rheumatism, eighteen cases cured.
Diseased kidney.
Asthma.
Neuralgia, four cases.
Enlarged glands, two cases.
Inflamed knee joint.
Contraction of an arm.
Headache, two cases of long standing.
Melancholy, two cases.
Deafness.
Withered arm after rheumatism.
Inflammation of the eye, two cases.
Quinsy.
Loss of voice.
Chlorosis.
Injured knee and arm, and sprained wrist.

1845.—Ophthalmia, four cases.
Inflammation of the lungs, and delirium; pronounced hopeless.
Rheumatism.
Neuralgia, seven cases.
Uterine disease.
Hysterical epilepsy, five cases.
St. Vitus's dance, two cases.

Nervous debility.
Insanity.
Brain fag.
Deafness.
Inflammatory loss of voice.
Erysipelas, two cases.
Contracted finger.
Diseased elbow.
Contracted foot.
Asthma, two cases.
Amaurosis.
Opacity of the cornea.
Stomach affection, with great emaciation.
Painless amputations, five cases.
Painless removal of tumors, three cases.
Painless removal of breast.
Painless removal of polypus from the nose.
Painless extension of a contracted cicatrix or scar
Several extractions of teeth without pain.

1846.—Epilepsy, nine cases.
Neuralgia, ten cases.
Rheumatism, thirteen cases.
Headache, six cases.
Loss of voice, four cases.
Fatuity and insanity.
Melancholia.
St. Vitus's dance.
Deafness, three cases.
Disease of spinal marrow.
Toothache, three cases.

Palsy of half the body.

Convulsions.

Brain irritability, severe.

Dumbness of many years standing.

Spasms.

Diseased breast, pronounced cancer and condemned to operation.

Diseased lung, pronounced hopeless consumption
by Dr. Todd.

Dropsy.

Inflammation of the eyes, four cases.

Disease of bladder.

Sprain.

Indigestion, two cases.

Amputations, six painless. cases.

Removal of tumors, twenty-one painless cases,
weighing from 30 to 112 lbs.

Cutting out a nail.

Application of strong acid to a sore.

Application of red hot iron to a sore.

Application of caustic to eye, many cases.

Removal, painlessly, of enlarged testis.

Removal of breast, two cases.

Tapping, two cases.

Operations for fistula, &c., eleven cases.

Injecting hydrocele, five cases.

Operation for cataract.

Extending a contracted knee.

Extractions of teeth, numerous.

Dividing the tendon of the heel.

1847.—Rheumatism, three cases.
Headache.
Hypochondriasis.
Rigidity of muscles.
Indigestion.
Irritation of the bladder.
Twisting of the head, two cases.
Affection of the heart.
Short-sightedness.
Ophthalmia, three cases.
Loss of voice.
Deafness, five cases.
Injury of the knee.
Abscess.
After pains.
Hiccough, 12 years' standing.
Strangury.
White swelling.
St. Vitus's dance.
Injury of the spine.
Pain of chest and spitting of blood.
Asthma.
Palsy.

Painless operations were—
Removal of a breast.
Removal of eight tumors, one 40 lbs. and another
100 lbs. in weight.
Opening of a whitlow.
Cutting down upon and tying an artery.
Extractions of teeth, numerous.

1848.—Neuralgia, eleven cases.
Rheumatism, sixteen cases.
Epilepsy, three cases.
St. Vitus's dance.
Loss of voice, two cases.
Painful affection of the heart.
Involuntary movements of an arm.
Cataleptic insanity.
Palsy, five cases.
Contraction of a finger.
Hysteria.
Chronic inflammation of an elbow.
Inflammatory swelling of the face.
Inflammation of the brain.
Inflammation of the throat.
Acute gout.
Headache, two cases.
Ulcers.
Lumbar abscess.
Diseased knee.
Enlargement of glands, three cases.
Painless operations through the year were :
Removal of tumors, some of very great weight and size, thirty-four cases.
Amputation of a great toe.
For hydrocele
Lithotomy.
Extraction of teeth, numerous.
Removal of scirrhus, two cases.
Caustic to sore.
Unconscious parturition, two cases.

1849.—Neuralgia cured, seven cases.

Rheumatism, thirty-six cases.

Palsy, five cases.

Epilepsy, two cases.

Insanity, five cases.

Deafness, two cases ; greatly improved five.

Dumbness from palsy.

Blindness of 26 years' duration, from opacity of the cornea ; also another of similar character of one year's standing.

Spinal affection of long standing.

Dropsy, two cases.

Palpitation.

Scrofulous ulcerations, much improved, four cases.

Nervous debility, three cases.

Vomiting, of two months' duration.

1850.—Cured—

Rheumatism, nineteen cases.

Hysteria, three cases.

Epilepsy, two cases.

Wry neck.

Gout.

Long standing vomiting.

Chronic headache, three cases.

Spinal irritation.

Neuralgia, ten cases.

Palpitation.

Spinal and general debility, two cases.

Nervousness.

Violent fits, two cases.

Deafness, three cases.
Insanity, three cases.
Delirium tremens.
St. Vitus's dance, three cases.
Loss of voice, five cases.
Inflammatory headache.
Diseased finger.
Spitting of blood.
Polypus of the uterus.
Ulcer of uterus, two cases.
Sprain.
Ulcerated glands.
Dropsy, two cases.
Chronic inflammation of lungs, two cases.
Ulcer in the neck.
Chlorosis, two obstinate cases.
Heart disease.
Inflammation of the eyes, five cases.
Erysipelas, two cases.
Inflammation of the knee-joint.
Inflammation of the bladder.
Delirium of fever.
Abscess of ear, of long standing.
Constipation, three cases.

1851.—The cures were—
Ulcers of the leg with varicose veins, two cases.
Chronic inflammation of the eye, with ulcers of
the cornea, two cases.
Injured leg.
Injured hip.

Lockjaw.
Chlorosis.
Chlorosis and neuralgia.
Deafness, two cases; one of seven years' duration.
Loss of voice.
Neuralgia, thirteen cases.
Wry neck.
Asthma.
St. Vitus's dance.
Convulsive and rigid fits.
Rheumatism, fifteen cases.
Epilepsy.
Spinal irritation, three cases.
Inflammation of the face.
Palsy of the legs.
Gout.
Dyspepsia.
Palsy, four cases.
Polypus of uterus.
Inflammation or neuralgia of uterus.
Insanity, two cases.
Hysteria, with convulsions and delirium.

1852.—Severe cutaneous disease.
Insanity.
Despondency.
Sleeplessness and dimness of sight.
Palsy, two cases.
Stiff knee.
Uterine disease.
Spinal curvature and hysteria.

Inflammation of the face.   ·

Rickets.

Erysipelas.

Epilepsy.

Gout.

Weakness of knees.

Weakness of ankles.

Spasmodic cough.

Vomiting blood.

St. Vitus's dance.

Neuralgia, nine cases.

Headache, five cases.

Loss of voice.

Palpitation.

Rheumatism, six cases.

Stiff knee.

Trembling and nervousness.

Blindness of seventeen years.

Lockjaw.

Inflammation of the eyes.

Removal of after pains.

The full particulars in each case of this great array of *solid facts* are carefully spread upon the record by Dr. Elliotson, and, coming from so eminent a source, require no endorsement from any man, and leave no chance for cavil at the immense power exerted by this great natural agent, for the cases themselves were generally of the most desperate and intractable character.

But I must be permitted to add, that many of

these statements have been abundantly verified in
my own experience during the past twenty-five
years. Some things stated may seem paradoxical;
it may be incomprehensible how a power, exerted
in the same way and apparently to the same
extent, shall at once be a "CALMANT and a STIMU-
LANT," and that the CIRCULATION by the same
means may be increased or depressed; but it is
nevertheless strictly true. I have often reduced
a pulsation from 145 to 100 in half an hour,
and found it remain at about this point up to
convalescence. I have raised a depressed circulation from an equally low point to its normal state
quite as quickly, and this is the uniform testimony
of all good operators. These effects are often
ABSOLUTELY under the WILL of the operator up to
the point where a natural equilibrium is reached,
and as they follow so directly and rapidly on the
initiation of the treatment, nothing is left to guess-
work: the results are seen at once, and, consequently, carry their own conviction.

In the line I have marked out for myself I omit
specific cases in my own practice, but they have
often been as startling as any I have seen recorded,
especially in spasmodic action.

I will, however, say this, in regard to my own
experience: There is *one* complaint prevalent in
many sections of this country known as Chills and
Fever, for which this treatment seems almost
specific. I have cured many cases of long standing
by a single treatment, often taking a case in the

midst of a violent rigor, and stopping the spasm in ten minutes—the cures remaining permanent. Little or nothing is published on this form of disease in foreign works on magnetism, and I call attention to it that others may make a similar test.

The first case of the kind I treated was my own brother, who had been suffering two months, with no progress towards recovery. He had been subjected to the usual heroic doses of quinine, arsenic, &c., without success. On the approach of a chill I one day magnetized him, and in ten minutes he was relieved; sleep supervened, which lasted thirty minutes longer, at the end of which he awoke, WELL, and has never had a symptom of its return, after a lapse of twenty-six years. He was, at the time, a lad of eleven years of age.

I could follow this up with many other remarkable cases, but this must suffice. I have only presented it here because of the personal knowledge of the *permanency* of the effects of a SINGLE treatment given a quarter of a century ago.

Sir J. D. Brandis, physician to the King of Sweden, cautious and conservative, could say that "magnetism will often be found efficacious in cases of defective sight and hearing, as also in CACHEXIOS of the vegetative organism—such as *scrofula*, rachitis, &c.—and more especially in those of the sensorial and intellectual systems." This eminent physician mentions the case of an elderly man who suffered from obstinate gout, with stiffness and calculous concretions in the joints, who was cured by the application of magnetism alone ; and also that of a young man affected with a severe paralysis of the leg and foot, with loss of flesh, and cold in the diseased limb, who was completely cured in the same way. (*See Brandis; ueber psychische Heilmittel und Magnetismus,* 1802.)

VII.

ANÆSTHETIC EFFECTS OF VITAL MAGNETISM.

THIS branch of the subject has always been received either with deadly opposition or open-mouthed wonder, and, since the application of ether, nitrous oxide and chloroform, it has been largely superseded by those agents ; their discovery, thirty years ago, alone prevented the general adoption, so far as applicable, of Animal Magnetism in painless surgery. Its introduction was greeted with bitter opposition by some of the great lights of the period, who solemnly contended that pain was a wise provision of nature, and should not be interfered with.

Dr. James Esdaile, surgeon in the hospitals of Bengal and Calcutta, in India, was most enterprising and successful in the application of magnetism in painless surgery. He published the results of his experiments in England in 1843, in a work entitled *Mesmerism in India*, in which he gives the particulars of two hundred and twenty-six cases, the majority of which were of a difficult nature. Amongst them were recorded amputations of breast and of arms; removals of fifteen tumors, weighing from eight to eighty pounds each; operations for stone; straightening contracted limbs, &c., &c.

These cases were perfectly authenticated by the best British residents of India, including the Governor-General, although the high character of Dr. Esdaile was itself a sufficient guarantee of the truthfulness of his statements. The publication of these facts produced an immense sensation throughout England, France and Germany, and they were speedily duplicated in various places on the continent of Europe. A detail of the experiments would fill a large volume.

In Dr. Brown Sequard's lectures upon "Nervous Force," delivered in Boston in 1874, he speaks of this form of anæsthesia as follows:

"As regards the power of producing anæsthesia, it seems to me unfortunate that the discovery of ether was made just when it was. It was, as you well know, in 1846 or 1847 that the use of ether as an anæsthetic was begun. It started from this city (Boston). At that time in England, Dr. Forbes was trying to show from facts observed in England, and especially in India, from the practice of Dr. Esdaile, that something which was called Mesmerism, but which, after all, was nothing but a peculiar state of somnambulism induced in patients, gave to them the idea that they were deprived of feeling ; so that they were in reality under the influence of their imagination, and operations were performed that were quite painless. I say that it was a pity that ether was introduced just then, as it prevented the progress of our knowledge as to this method of producing anæsthesia. My friend Dr. Broca took it up in 1857–8 and pushed it very far; and for a time it was the fashion in Paris to have amputations performed after having been anæsthetized by the influence of Braidism or Hypnotism. A great many operations were performed in that way that were quite painless. But it was a process that was long and tedious, and surgeons were in a hurry and gave it up.

"I regret it very much, *as there has never been a case of death from that method of producing anæsthesia,* while you well know that a great many cases of death have been produced by other methods.

"Not only anæsthesia may be produced, but the secretions may be very powerfully affected by the influence of the mind over the body.

"Every one knows, also, that the secretions of bile, the secretions of tears, and the secretion of saliva, are very much under the influence of the nervous system. The purging of the bowels, which depends on a secretion there, or a secretion in the liver, is also much dependent on the influence of the imagination." (*N. Y. Tribune Extra, April* 9, 1874, *p.* 34.)

In reply to the very natural inquiry why this SAFE method has not been adhered to, the remark of Dr. Sequard affords the only solution I know of, viz.: *"It was a process that was long and tedious, and the surgeons were in a hurry and gave it up."*

It would seem, however, that most people who desire to avoid the risk of DEATH would try to obtain this desirable result, so concisely summed up by Dr. Sequard: *"There has never been a case of death from that method of producing anæsthesia, while you well know that a great many cases of death have been produced by other methods."*

For GENERAL practice I admit that the excuse of the surgeons "being in a hurry" has force, but we can imagine many cases where anxious friends would be in a position to make the surgeons wait until this method had been tried, before incurring ANY risk that could otherwise be avoided.

VIII.

If nervous force, or vital magnetism, could be bottled up and purchased of the apothecaries, it would at once revolutionize the practice of medicine; but in a practice that needs a grand natural adaptation, moral and physical qualities of the highest order, and a benevolence untiring under labors the most exhausting, we can see difficulties mountain high to embarrass physicians and surgeons described by Dr. Sequard as being "in a hurry."

In no one thing do writers upon magnetism agree more perfectly than in requiring the very highest personal and moral qualifications.

Without enumerating authors *in extenso*, I will group a few of these requirements as set forth by others.

"Benevolence of disposition; a strong, steady will to direct; power of abstraction and concentration; confidence in himself without vanity; established moral character; well regulated life; delicacy of manner; patience and good temper."

In connection with these qualities, must be, first and controlling, an inborn natural magnetic force, able to expend itself without material exhaustion;

and thoroughly successful magnetic practitioners are found perhaps in the ratio of only one in a thousand persons.

It may be objected to this practice that its demands are singular and impracticable; but a little reflection will show that they are paralleled in other departments of the healing art. Doctor Henry A. Hartt lectured in Association Hall, New York, recently, in favor of the establishment of a "Hospital for the treatment of Chronic Diseases." After a searching analysis of past and present methods of treating disease, he made the following declaration in regard to the qualifications of practitioners for the sphere of labor which he was advocating:

"The treatment of chronic diseases requires a peculiar mental and moral constitution, and it would be just as impossible to find in every physician a man adapted to this particular field, as it would be to find in every clergyman a brilliant orator, in every lawyer an able jurist, in every scientist a profound logician, or in every student of literature a magnificent poet. The grand qualifications for this sphere of labor are *grit, patience, energy, and the irresistible magnetism of an unconquerable faith.*"

Words fitly spoken; true of every man who really succeeds as a physician; and without these endowments no amount of technical knowledge will bring brilliant success.

But, crucial as this test might appear as applied to the flippant "younger members of the Association" who received Dr. Hartt's suggestions with "marked dissatisfaction," the silvered heads of experience, embracing so many of the wisest men in the profession, thoroughly endorsed his views. No better statement of the qualifications of a successful *magnetic* physician could have been made than that so tersely summed up in *"grit, patience, energy, and the irresistible magnetism of an unconquerable faith;"* for these elements cover that "mental and moral constitution" so eminently required in a good magnetic practitioner; and they are just as rare in the field of magnetic practice as in the medical world. These qualities are more imperatively demanded in magnetic practice, because, while most forms of ACUTE attacks flee readily before the magnetic touch, it is generally in chronic forms of disease, when medication has done its best, or its worst, when hope has well nigh taken its eternal flight, and the vital spark is almost extinguished, that magnetism is sought as a last resort, and a miracle expected. In such cases faith has again to be rekindled, hope inspired, and again and again, untiringly, a new nervous life is required to be poured upon the patient, until he has reached his own normal nervous equilibrium.

To calmly, patiently and faithfully respond to a demand like this, so exhaustive in its character, requires, in the operator, heroic qualities of benevolence, demanded of no other practice, for he is parting with a vitality all his own, which no mediumistic theory, or ghostly supernatural intervention will make good. Nature's divine law of compensation can alone do this. Every one who has trodden the path will testify to this truth, and the experienced alone are competent to speak thereon.

The mere mechanical attitude of the magnetic physician will never achieve noble results; while it may produce a certain class of phenomena upon a small percentage of *natural sensitives*. For the overthrow of obstinate forms of disease there is needed a vitalizing energy, that no RUBBING process, MASSAGE, *tripsis*, frictions or mechanical MOVEMENT CURE can impart; and valuable as are these processes for nerve exercise, they must not be confounded with vital magnetism, as is often done.

I have known intelligent physicians to recommend a "rubbing doctor" for cases where REST was required, and not MOTION; the consequence was a restless, sleepless, nervous condition, instead of the soothing influence which magnetism would produce. They charged the failure of the effort to magnetism, which, *in their ignorance*, they supposed they were prescribing.

Only when magnetism is as well understood here as it is in Europe, and when it has been as carefully studied in its effects, will errors be avoided both in the professional and non-professional mind.

In France it is well established as a practice. In Prussia it is used extensively by the authority of government. In Sweden degrees are granted by the University of Stockholm to those who stand an examination upon its laws and effects. In Russia, a commission of médical men, under direction of the Emperor, inquired into it thirty years ago, and at that time reported it a " very important agent," and the first physician of the Emperor, with other eminent Russian physicians of St. Petersburgh, commended in highest terms its utility; and at Moscow a systematic course of treatment, under the best auspices, has been employed for years.

In Denmark it is practiced by physicians under a royal ordinance, and by a decree of the College of Health. It has made the same advances in Holland; and has numbered a host of eminent English physicians amongst its friends for the last thirty years.

A recently returned lady missionary from Burmah, who comes home after an absence of ten years, informs me that it is extensively and intelligently practiced amongst the Burmans; that she herself had received the best effects from the treatment in a nervous prostration induced by the enervating climate, when every other means had failed.

The natives denominate it " The medicine over all medicines."

IX.

It was a fashion for a time to speak of magnetic cures as some form of delusion, and in the midst of the war upon the science and its facts, thirty years ago, Miss Harriet Martineau, the celebrated English authoress, was restored to health by magnetic treatment after an exhausting illness of five years' duration.

The following account by this gifted lady was published in a series of letters, in 1845, and gives a clear and lucid statement of an experience very touching and convincing, and marked with the same vigor that accompanied everything that emanated from her pen.

She says: "During these five years, I never felt wholly at ease for a single hour.

"I seldom had severe pain; but never comfort. A besetting sickness, almost disabling me from taking food for two years, brought me very low, and, together with other evils, it confined me to a condition of almost entire stillness—to a life passed between my bed and my sofa.

"It was not until after many attempts at gentle exercise that my friends agreed with me that the cost was too great for any advantage gained; and at length it was clear that even going down one flight of stairs was imprudent.

"From that time I lay still; and by means of this undisturbed quiet, and such an increase of opiates as kept down my most urgent discomforts, I passed the last two years with less suffering than the three preceding.

"There was, however, no favorable change in the disease. Everything was done for me that the best medical skill and science could suggest, and the most indefatigable humanity and family affection devise; but nothing could avail beyond mere alleviation. My dependence on opiates was desperate. My kind and vigilant medical friend,—the most sanguine man I know, and the most bent upon keeping his patients hopeful,—avowed to me last Christmas, and twice afterwards, that he found himself compelled to give up all hope of affecting the disease,—of doing more than keeping me up, in collateral respects, to the highest practicable point. This was no surprise to me; for when any specific medicine is taken for above two years without affecting the disease, there is no more ground for hope in reason than in feeling.

"In June last, I suffered more than usual, and new measures of alleviation were resorted to. As to all essential points of the disease, I was never lower than immediately before I made the trial of mesmerism.

"If, at any time during my illness, I had been asked with serious purpose, whether I believed there was no resource for me, I should have replied that mesmerism might perhaps give me partial relief. I thought it right, and still think it was right, to wear out all other means first. It was not, however, for the reason that the testimony might be thus rendered wholly unquestionable— though I now feel that my years of suffering were but a light loss for such a result;—it was for a more personal reason that I waited.

"Surrounded as I was by relations and friends, who, knowing nothing of mesmerism, regarded it as a delusion or an imposture,—tenderly guarded and cared for as I was by those who so thought, and who went even further than myself in deference to the ordinary medical science and practice, it was morally impossible for me to entertain the idea of trying mesmerism while any hope was cherished from other means."

After having decided to make a trial of vital magnetism, and the lapse of some time, she proceeds:

"At the end of four months I was, so far as my own feelings could be any warrant, quite well. My mesmerist and I are not so precipitate as to conclude my disease yet extirpated, and my health restored beyond all danger of a relapse; because time can only prove such facts. We have not yet discontinued the mesmeric treatment, and I have not yet re-entered upon the hurry and bustle of the world. The case is thus not complete enough for a professional statement.

"But, as I am aware of no ailment, and am restored to the full enjoyment of active days, and
nights of rest, to the full use of my powers of
body and mind; and, as many invalids, still lan-
guishing in such illness as I have recovered from,
are looking to me for guidance in the pursuit of
health by this means, I think it right not to delay
giving a precise statement of my own mesmeric
experience, and my observations of some different
manifestations in the instance of another patient
in the same house.

" A further reason against delay is, that it would
be a pity to omit the record of some of the fresh
feelings and immature ideas which attend an early
experience of mesmeric influence, and which it may
be an aid and comfort to novices to recognize from
my record.

" And again, as there is no saying, in regard to a
subject so obscure, what is trivial and what is not,
the fullest detail is likely to be the wisest, and the
earlier the narrative the fuller; while better knowledge will teach us hereafter what are the non-essen-
tials that may be dismissed.

" Nothing is to me more unquestionable and more
striking about this influence than the *absence of all
reaction*. Its highest exhilaration is followed, not by
depression or exhaustion, but by a further renovation. From the first hour to the present, I have
never fallen back a single step. Every point gained
has been steadily held.

" Improved composure of nerve and spirits has

followed upon every mesmeric exhilaration. I have been spared all the weakness of convalescence, and carried through all the usually formidable enterprises of return from deep disease to health, with a steadiness and tranquility astonishing to all witnesses.

" At this time, before venturing to speak of my health as established, I believe myself more firm in nerve, more calm and steady in mind and spirits than at any time in my life before.

" So much consideration of the natural and common fear of the mesmeric influence as pernicious excitement, as a kind of intoxication.

" On four days, scattered throughout six weeks, our séance was prevented by visitors or other accidents. On these four days the old distress and pain recurred: but never on the days when I was mesmerized.

" From the middle of August (after I had discontinued all medicines but opiates) the departure of the worst pains and oppressions of my disease made me suspect that the complaint itself, the incurable, hopeless disease of so many years, was reached: and now I first began to glance towards the thought of a recovery.

" In two or three weeks more it became certain that I was not deceived; and the radical amendment has since gone on without intermission."

At this time a change was made in the magnetizer for one of higher power, and she continues :

" The visual appearances were much the same as before, but the experience of recovery was more rapid. I can describe it only by saying, that I felt as if my life was fed day by day. The vital force infused or induced was as clear and certain as the strength given by food to those who are faint from hunger.

" I am careful to avoid theorizing at present on a subject which has not yet furnished me with a sufficiency of facts; but it can hardly be called theorizing to say (while silent as to the nature of the agency) that the principle of life itself—that principle which is antagonistic to disease—appears to be fortified by the mesmeric influence; and thus far we may account for mesmerism being no specific, but successful through the widest range of diseases that are not hereditary, and have not caused disorganization.

" No mistake about mesmerism is more prevalent than the supposition that it can avail only in nervous diseases.

" The numerous cases recorded of cure of rheumatism, dropsy, cancer and the whole class of tumors— cases as distinct and almost as numerous as those of cure of paralysis, epilepsy and other diseases of the brain and nerves—must make an inquirer cautious of limiting his anticipations and experiments by any theory of exclusive action on the nervous system. Whether mesmerism, and, indeed, any influence whatever, acts exclusively through the nervous system is another question."

Her abandonment of opiates is then described in the final happy termination of her case :

"The same fortifying influence carried me through the greatest effort of all—the final severance from opiates.

"What that struggle is, can be conceived only by those who have experienced, or watched it with solicitude in a case of desperate dependence on them for years. No previous reduction can bridge over the chasm which separates an opiated from the natural state.

" I see in my own experience a consoling promise for the diseased, and also for the intemperate, who may desire to regain a natural condition, but might fail through bodily suffering. When the mesmeric condition can be induced, the transition may be made comparatively easy. It appears, however, that opiates are a great hindrance to the production of the sleep; but even so, the mesmeric influence is an inestimable help, as I can testify."

Miss Martineau sums up the account of her own case in the following manner:

"Before leaving the narrative of my own case for that of another, widely different, I put in a claim for my experiment being considered rational. It surely was so, not only on account of my previous knowledge of facts, and of my hoplessness from any other resource, but on grounds which other sufferers may share with me:—on the ground that though the science of medicine may be exhausted in any particular case, it does not follow that curative means

are exhausted:—on the ground of the ignorance of
all men of the nature and extent of the reparative
power which lies under our hand, and which is
vaguely indicated under the term " Nature ":—on
the ground of the ignorance of all men, regarding the
very structure, and, much more, the functions of
the nervous system:—and on the broad and ultimate
ground of our total ignorance of the principle of
life—of what it is and where it resides, and whether it can be reached, and in any way beneficially
affected by a voluntary application of human
energy.

" It seemed to me rational to seek a way of re-
freshment first, and then to health, amidst this
wilderness of ignorance, rather than to lie perishing
in their depths.

" The event seems to prove it so. The story appears to me to speak for itself. If it does not
assert itself to all,—if any should, as is common in
cases of restoration by Mesmerism, try to account
for the result by any means but those which are
obvious, supposing a host of moral impossibilities
rather than admit a plain new fact, I have no con-
cern with such objectors or objections.

" In a case of blindness cured, once upon a time,
and cavilled at and denied, from hostility to the
means, an answer was given which we are wont to
consider sufficiently satisfactory, ' One thing I know,
that whereas I was blind, now I see.' Those who
could dispute the fact after this must be left to
their doubts.

"They could, it is true, cast out their restored brother; but they could not impair his joy in his new blessing, nor despoil him of his far higher privileges of belief in and allegiance to his benefactor.

"Thus, whenever, under the Providence which leads on our race to knowledge and power, any new blessing of healing arises, it is little to one who enjoys it what disputes are caused among observers.

"To him the privilege is clear and substantial. Physically, having been diseased, he is now well. Intellectually, having been blind, he now sees. For the wisest this is enough. And for those of a somewhat lower order, who have a restless craving for human sympathy in their recovered relish of life, there is almost a certainty, that somewhere near them there exist hearts susceptible of simple faith in the unexplored powers of nature, and minds capable of an ingenuous recognition of plain fact, though they be new, and must wait for a theoretical solution."

These are not the words of a deluded, weak-minded enthusiast; they are the strong clear words of one of the most gifted of England's daughters, ringing out at a time when envy, jealousy, and distrust, assailed the possibility of magnetic cure; and so valuable was this testimony in the stormy days of this science that the heroic sentiment of Miss Martineau seems vindicated when she asserts that her "years of suffering were but a light cost" that her testimony "might thus be rendered wholly unquestionable." In passing judgment on her

statement we may well recall the high mental cul·
ture of the witness as displayed in her productions,
whose power and versatility establishes the value
of her personal testimony in favor of Magnetism,
while her refusal to accept a pension from the
British Government, when twice tendered, in ac-
knowledgment of her services as a writer on questions of political economy, because it would conflict with the spirit of her own teaching, guaran-
tees an independence and honesty of expression
beyond cavil.

This esteemed woman's life had its burden lifted,
and was prolonged thirty years by the application
of vital magnetism, and the world has been made
so much richer by the contributions of the brain of
one of its best and sweetest workers.

The full particulars of the uterine disease under
which Miss Martineau suffered, will be found detailed by Doctors Naido of Venice, and Greenhow,
and Sir C. M. Clarke, of England, in vol. 3, pp. 89,
90, 91—2, of the *Zoist.*

The grossest misrepresentations were made, both
as to the cure itself and Miss M.'s views upon the
subject : and to correct these she wrote as follows :

AMBLESIDE, Dec. 11, 1845.

SIR : I am greatly obliged by you letting me know, through
Mrs. ———, the notions that are abroad on the subject of
my health and my present views of mesmerism. Nothing can
be more absurdly false than they are. I am in robust health,
and have not had one day's illness since I avowed my cure by
mesmerism. My long daily walks and the literary labor I un-
dergo without fatigue, have satisfied all who know me that I

am perfectly well. I myself am fully aware that I am well for the first time in my life, and that I owe my health wholly and solely to mesmerism. My gratitude to those who guided me to this remedy, and who administered it is, I need not say, as strong as in the early days of my recovered case.

That my convictions remain what they were a year ago, is known to all my personal acquaintances, because they are aware that if I had changed my opinions, I should have made a recantation as free and full as my avowal.

I trust, too, that they know that a fact in natural philosophy, once ascertained by experiment, can never again be disbelieved.

But, as there are many persons who know neither of these things, and who yet may have power to discourage inquiries into mesmerism, an inquiry which I think it my duty to promote by any means in my power, I have no objection to authorize your making any use you think proper of this letter.

I remain, sir, sincerely yours,

HARRIET MARTINEAU.

Capt. James.

The grand tributes of admiration bestowed by so many able writers upon the genius of this distinguished woman since her decease, bring all her graces and powers to mind, and remind us that the memory of the just is precious.

THE EXTRAORDINARY CASE OF ANNE VIALS.

Rev. Geo. Sandby, in his valuable work "Mesmerism and Its Opponents," published in London, gives a circumstantial account of a most extraordinary case as follows ;

"Anne Vials is the daughter of Samuel Vials, of the Albany Parish in St. Alban's, who formerly drove the mail cart from thence to Watford. For a short time this poor girl gained her livelihood by working in a silk factory. From the scrofulous character of her constitution, she was not always equal to full employment ; but in 1837, when she was only sixteen years of age, she was compelled to give up work altogether. For her mother fell sick with a long and pining illness, under which, after much suffering, she finally sank ; and during which she was confined to her bed, and required the constant presence of a nurse.

" Poor Anne, therefore, left her calling at the factory, took her place at her mother's couch, and was her unwearied attendant night and day. So feeble, indeed, was the patient, that she could scarcely be quitted for a moment ; and for a long year, therefore, did this anxious and affectionate child sit by her parent's bed the whole night through. When death at length released the sufferer a fatal discovery was made. The mother's disease had taken strong hold of the daughter, for the overwrought exertions of a twelvemonth had now too clearly brought out the hereditary taint.

"Anne Vials, in fact, required a nurse herself ; for not only was the general state of her health broken down, but the left arm, which for three or four years had been giving her much pain and uneasiness, became now so diseased as to totally deprive her of its use. She was placed under the care of several medical men in succession ; the best attendance in St. Alban's was provided for her ; but the arm every day grew more and more painful. Through the kindness of some charitable friends she was now admitted into different hospitals, one after the other. She was first removed to Hewel Hempstead Infirmary, thence to St. Bartholomew's Hospital, London, where she remained nine months ; thence to St. Thomas's, in the Borough, and thence to Hewel Hempstead again, in none of which did she obtain any effectual benefit. The state of her health at length became so serious that to save her life some decisive measures were necessary, and she was taken up to London again, to Guy's Hospital, where her arm was amputated by Mr. Morgan the 22d of March, 1841.

"At the end of three months, when the wound was healed, she returned to St. Alban's. After she had been at home some little time, a violent convulsive action commenced in the stump. This movement grew rapidly worse and worse. In fact, the stump moved up and down, day and night, unceasingly,—and much quicker, to use her own expression, than she herself could move the other arm.

" Her sufferings became intense, and her general health was affected in proportion. She was now moved backwards and forwards, as before, to the different hospitals, but without any relief. At the infirmary in Hewel Hempstead, they actually strapped the arm down, with the hope of lessening the movement ; but the confinement, if possible, made it worse, and they were compelled to unloose it.

" She was at length carried to St. George's Hospital; here she remained three months : her health
gradually getting worse, and the epileptic fits, from
which she had been suffering for a twelve-month,
increasing in violence and duration ; when, with
the only hope of saving her life a proposition was
mooted of taking the stump out of the socket. My
readers may judge by this simple fact, of the desperate state to which this poor girl had now arrived ;
for with her shattered health, it could hardly be
expected that she should survive, even for a short
time, so serious an operation. Fortunately for poor
Anne, she had several benevolent friends, who,
knowing all the circumstances of her history, had
watched the fearful progress of her sufferings from
the first ; and by subscriptions and various little
Christian kindnesses had done much towards lessening her load of sorrow ; Mr. Basil Montague, in
particular, that excellent man, whose long and useful life has been devoted to the benefit of his fellow
creatures, took the warmest interest in her fate ;
she often went to his house, and there she received
from Mrs. Montague that sympathy and consideration which woman alone is able to bestow.

"One day the thought struck both these kind
friends, that if anything could be of service to
Anne in this extremity of misery, it was Mesmerism.

" It was the faintest hope, for they had but slight
knowledge or belief of its power :—still they mentioned the case to their friend, Mr. Atkinson, and

suggested to him the idea of making a trial of what could be done.

"In spite of the feeling against Mesmerism, towards him, and the almost hopeless state of the patient, he at once on his own responsibility undertook the case; and seeing that it would require, for months, the most unremitting attention, he procured a nurse from St. George's Hospital, and had the poor girl removed to his own house.

" It was in May, 1842, about fourteen months after the amputation, that Anne Vials quitted the hospital to make trial of Mesmerism; and this is the description of Mr. Atkinson at the time he found her : She had sometimes three or four fits in a day, of a most violent nature, which continued for more than an hour; the stump moved up and down without cessation, not a merely nervous twitching, but violently up and down; she suffered continuous excruciating pain in the head and back, and at the top of the stump too the pain was most excruciating; she had a pain too in all her limbs and joints, particularly in the elbow of the remaining arm, just as she had before amputation in the other.

" Masses of sores were constantly breaking out in different parts of the body; palpitations at the heart, pains in the chest, suspension of the functions of nature, and a spitting of large quantities of blood accompanied by solid matter, were some of the other symptoms.

" In short, a more terrible complication of evils have seldom been united in one sufferer.

" I shall leave it to Mr. Atkinson to give at some
future period to the public the interesting details
of his success.

" Let it be sufficient to state that the process was
most painfully laborious, occupying a large portion
of his time, and that she remained in his house
more than twelve months.

" At the first few sittings the epileptic fits were
brought on, as if by the Mesmeric effect; but this
prevented their recurrence in her ordinary state.
At the fourth SEANCE, the deep sleep or trance was
superinduced, *when the action of the stump suddenly
stopped*, and from that time it never moved in that
way again; the fits, too, ceased; the pains in the
back of her head were almost immediately relieved;
and a gradual improvement in her general health
set in.

" Upon the wonderful results of Mesmeric treatment in this case I shall make little comment; my
readers can think for themselves; they will see
here a poor girl, carried to and fro from hospital
to hospital, enduring the most exquisite torture,
and her life placed in such a state of jeopardy that
the only hope of preserving it was recourse to a
second and horrible operation. The arm was to
be taken out of the socket ! An effectual mode, in
truth, for a prevention of its movement !

" But from this operation was she spared by the
action of Mesmerism; by its continued and regular
application was a relapse prevented and an improvement in her health obtained.

"Who does not see the goodness of Providence in vouchsafing such an agent? Who can deny that Mesmerism was to her the precious gift of God?

"The facts of her case,—of her sufferings, of the amputation, of the movement of the stump, and of the other attendant evils, are known to numbers,—to medical men in St. Alban's,—and to the surgeons and nurses in the hospitals; and it is also known, that all the remedies suggested for her benefit were fruitless; the best surgical advice was of no avail; but the fifth day, after the application of Mesmerism, the stump ceased to move, and the other fearful symptoms began to disappear."

Mr. Sandby gives a very interesting account of curious and beautiful mental phenomena which attended this case in its progress towards a cure, but I have confined myself to the physical effects, as they alone come within the scope of my present purpose.

A DOUBLE PHENOMENON.

A very notable case is referred to by Rev. C. H. Townshend, in his "Facts in Mesmerism." He says: "Chardel, a French physician, and writer on magnetism, gives an interesting account of two sisters whom he mesmerized as a physician, with the hope of checking a tendency to consumption, which they had both evinced. One evening, being in *sleep-walking*, they, as if prompted by natural instinct, entreated their mesmerizer to leave them in that state, only so far de-mesmerizing them as to open their eyes, and to be committed to their own self-guidance. Day after day they renewed their petition,—for day after day they felt health returning under the magnetic influence. In other respects they pursued their usual habits, and their mesmeric existence had its alternate periods of sleep and of waking, as regular as those of the natural life.

"At the end of three months, their case appearing to be complete, M. Chardel conducted the sisters, accompanied by their mother, to a beautiful spot in the country, where he restored them to a knowledge of themselves. He describes, in lively terms, their surprise and joy at returning consciousness.

" It was winter when they entered the mesmeric state—it was now spring. The ground was then covered with snow, but' now with flowers. They were then looking forward to an early grave; but now the feeling of renewed health tinged everything with hope and life. Almost doubting if they did not dream, they threw themselves into their mother's arms, gathered flowers, and smelt them, and endeavored, by the exercise of their senses, to convince themselves that it was all a blessed reality. Not a circumstance of the past three months survived in their memory."

Here we have a case of physical and psychological phenomena united, which must be my apology, if any is needed, for introducing it.

CASE OF UTERINE DISEASE WITH GENERAL NERVOUS DERANGEMENT.

Doctor Henry Storer, of Queen Adelaide's Hospital, London, reports an interesting case in his work, " Mesmerism in Disease," that was treated by Mr. Vernon, of the " Institute of Mesmerism," London. The patient writes:

" *Sir:* At Mr. Vernon's request I send you some account of the good resulting from the application of mesmerism in my own case, after an illness of five years. I have suffered from an internal complaint of a very painful and aggravated nature, partly originating from an adverse circumstance at the birth of an only child.

" I experienced little or no relief from surgical aid during the period I have mentioned. A detailed account of the many years during which my health has slowly but surely declined, would be unnecessary. I will but briefly explain my state when forced to apply to mesmerism for relief, as my last resort. I unfortunately met with a severe accident in the beginning of last March, in consequence of which my complaint was so much aggravated, that in the course of a few weeks I was entirely confined to the sofa.

"I was attended by Mr. R. Taylor, surgeon, of Brighton, and Dr. G. Hall, of 14 Old Stein. Mr. Taylor's reputation is too high in Brighton for me to doubt that his treatment of me was otherwise than judicious. I believe that all was done for me that surgical skill could do, notwithstanding which, each week, almost each day, saw me worse, until I was unable to walk, or even move, without the greatest pain. From June last until the end of November I was leeched every fourth day, yet I experienced little or no relief from this, or any other treatment; my sufferings were incessant and agonizing.

"I was only enabled to bear them by frequent and large doses of morphine and other sedatives, which, together with the pain, produced an equally distressing effect on the mind. My mother was frequently told by Mr. Taylor *that he knew not what to do for me ;* that I must trust to time and to the advantages which I received from my age (25).

" *This declaration was repeated only two days before I was mesmerized,* when I was confined to my bed, and felt my strength failing rapidly, so that I could not endure much longer.

"I freely confess that had I not been so hopeless of relief, I *would not* have applied to mesmerism, from the fact of the prejudice attending it. I will only premise that I believed in the *power*, but I did not believe the possibility of its efficacy in my case.

"On the evening of the 4th of December, 1844, Mr. Vernon mesmerized me. When he arrived I was in extreme pain, and in a state of excessive irritability and restlessness, so that I did not believe it possible for me to lie still to allow Mr. V. to make the attempt; notwithstanding which, after what appeared to me to be a very short time, I felt that the pains moderated, and a sensation of ease and freedom from restlessness steal over me, which no opiates could ever induce. I soon lost all consciousness, and on awaking was almost free from pain.

"That night I was far calmer than I had been for many months, although the pains returned.

"Mr. Vernon continued to mesmerize me twice a day, with one or two exceptions. At the end of a week I found myself enjoying several hours of quiet and refreshing sleep at nights. After a fortnight's mesmerizing, I discontinued the opiates which I had regularly taken for many months, and was able to walk about the room without much inconvenience ; and at the end of three weeks I found sleep, appetite, and strength of body and mind restored to a degree which astonished no one more than myself.

"My mother, by whose advice and persuasion I was induced to try mesmerism, removed with me to London in order to perfect my cure, which object has been most happily and decidedly accomplished, for I am enabled to employ and exert myself as well as I did six years ago, and can take out-door exercise without fatigue and inconvenience. As a proof of which, I walked three miles on Friday last with perfect comfort to myself.

" Should you, or any other medical man, wish for any more detailed account of my illness and cure, I must refer you to Mr. Vernon, who, I am sure, will be happy to reply to any inquiries.

" Anything in my power to further so good and powerful a cause as mesmerism, will be gladly complied with."

Doctor Storer, of Bath, is sufficient authority for the introduction of the above statement.

CASE OF NERVOUS SUFFERING LEAVING THE PATIENT
DUMB.

The following case from the *Zoist*, is interesting, having occurred in Devonshire, and been recorded in the *North Devon Journal :*

" Many of our readers will be gratified to know that thanks were publicly returned in the Parish Church of Ilfracombe, on Sunday last, for the signal mercy of God toward the girl, Catharine Brown, in her restoration of speech by mesmerism, through the instrumentality of Mr. Davey.

"'The girl is in her fifteenth year; her name, Catharine Brown ; she is the daughter of a mechanic living at Compass Hill, Ilfracombe.

" Her mother states that she was taken alarmingly ill in the month of October, 1841 ; her complaint lying in her head, side and stomach.

" At this time she presented a most deplorable picture of human wretchedness and suffering, and her screams, which were terrific, could be heard a long distance.

"She could not be left for a moment, as, during the absence of the attendants, she would beat her head against a wall most unmercifully, so as to leave wounds and bruises. She was likewise often convulsed, and, during the paroxysms, it was as much as two could do to hold her.

"Many medical men of the town visited her, and gave her medicine ; in fact, she was an object of general charity, and, I believe, there was not even an occasional medical visitor in the town who did not prescribe for her.

"After a lengthened period of suffering, the violence of her complaint greatly subsided, but *left her dumb !* From that period up to the time of being first mesmerized, she had never spoken a single syllable.

"After being mesmerized three times, and as soon as consciousness was suspended, she began to speak. At first her speech was slow, dwelling upon the first letter of the word, as in stammering, after a while she could distinctly pronounce her name, 'Catharine Brown.' Those who were present will not soon forget the countenance of the patient upon being de-mesmerized. Rising from her chair, she, ran towards a little girl in the room, crying out in joyous accents, 'Oh, Mary Ann, I can speak now!'

"A fear was entertained that the faculty of speech would again be lost when the patient was de-mesmerized ; but the fear was groundless, and was wholly dissipated, on hearing her thus address her little friend.

"The delight of the mother may be more easily imagined than described, when she found her daughter able to speak, *after being dumb for the space of two years and a half.*"

————

CURE OF DEAFNESS OF LONG STANDING.

Dr. Storer reports a case that was treated by Mr. Saunders, and came under his observation, as follows : "Mary Ann Parsons, age 24, residing at Combe Down, near Bath, has been afflicted from her early youth with deafness. Her mother, having been relieved of rheumatism by mesmerism, brought her daughter for the purpose of being mesmerized for deafness.

"In the mesmeric trance she prescribed 'breathing in each ear.' Within the last month we have mesmerized her fifteen or sixteen times, and her deafness is entirely removed.

"She at last became so susceptible, that two or three passes sufficed to send her off into the deepest coma.

"Before mesmerism she could not hear the ticking of a watch when close to her ears ; now she can hear a loud ticking clock at the distance of a second room, and continues to this period quite well.

CASE OF SPASMODIC ASTHMA.

Dr. Storer records another case treated under his supervision by Mr. Kiste. The patient was Elizabeth Spardens, the wife of the footman of the Rev. Mr. Holdsworth, who writes as follows:

"According to your wish, I send you a copy of the notes taken at the time when you favored us with your company, and mesmerized the wife of my footman—an operation which has so far been of most essential service to her.

" The asthma, under which she has so long suffered, was of a most distressing character ; so much so, that her neighbor, who inhabits a part of the house, was afraid to remain with her alone in the house unless her husband or some friend of hers was there also, so frightful was her appearance when under the attack. You know already that for six weeks after she was put under mesmeric influence she had no attack ; then a slight one, brought on, she says, by her own imprudence, and from that time, now six weeks more, she has had no return of her complaint ; her general health is improved, and she is gaining more solid flesh. I hope that in time the beneficial effects of mesmerism will be better understood and better appreciated. Nothing can be more unphilosophical than the manner in which the faculty and others treat this subject, as well as the prejudices that have been evinced, instead of the patient investigation which such a subject requires.

"If doubt leads us to neglect investigation, and prejudice prevents inquiry, how is knowledge in any science or subject to be acquired ?

"BRIXHAM VICARAGE, Feb. 5, 1841."

Dr. Storer says the case had been under the care of several medical men in the locality, who pronounced her case incurable, and of the worst possible character.

The daily details of the treatment are too long for insertion ; a year has now elapsed, and she has enjoyed uniformly good health. Seeing me lately, she said in her simplicity, that frequently when sitting by herself she would "look back to former years, and fancy she could not be the same person."

———

CASE OF NEURALGIA OR PAINFUL NERVOUS AFFECTION OF THE HEART OF MANY YEARS STANDING.

Dr. Storer reports another very interesting case, Mr. Kiste again being the operator, and the Hon. Mrs. Hare the subject.

For EIGHTEEN years she has suffered from neuralgia of the heart, and was successively under the care of Dr. Warren, Sir C. Bell, Mr. Abernethy, Dr. Maton, and Dr. Elliottson, but the various remedies which were suggested did not in the least relieve her.

At this period she was taking *forty* grains of opium a day. Her sufferings continued unalleviated till the end of last year, (1844,) when she was fortunately introduced to Mr. Kiste. On the 16th of Sept. last, the lady writes, after describing her doubts as to the probable effect :

"In a few minutes I felt a most unusual tranquility of mind ; the objects in the room seemed to lose their outline, and the last things I remember to have seen were my mesmerizer's eyes.

"I was afterwards told that in eight minutes I sank down in the arm-chair, and that Mr. K. pronounced me to be in what is called mesmeric somnambulism, a peculiar state of the nervous system.

"After nearly two hours' sleep Mr. K. awoke me, and I felt tranquil, but very sleepy.

"That night, for the first time, after having had paroxysms every night for three weeks previously, I was not only free from an attack, but I slept soundly, till late in the morning, when I awoke quite refreshed.

"Since that day—Sept. 17—I have been entirely free from these attacks, with the exception of several times I felt the symptoms when I had taken cold, or when Mr. K. was not with us ; but it has not come to anything more than a little faintness.

"*I know nothing of my former agonies, except by memory.*

"My general constitution has changed. I am much thinner ; and in the course of three months I have twice had occasion to have my dresses much diminished, and I am now able to walk five or six miles without the least fatigue. Having been for nearly two months free from my usual sufferings, Mr. Kiste proposed to me to diminish my doses. Although I thought this an impossibility, I made the trial, when I found I did not, as formerly, feel the loss.

"I continued by degrees to lessen the quantity, and from having at that time been in the habit of taking from sixty-six to eighty-six grains of opium daily, I have now reduced it to less than thirty, and am going on to diminish the quantity.

"I was a dreadful sufferer for many years; I now am free from that suffering, and in the enjoyment of health and comfort. Whereas I was ill, I am now well! In thus coming forward, I am fully aware that I expose myself to the ridicule of small minds and the observations of the malevolent; to both I am perfectly indifferent."

———

CASE OF SEVERE NERVOUS DISEASE.

Dr. Storer reports the case of a lady named Larke, the daughter of a medical gentleman residing and practising at Norwich; her case was described as deplorable in the extreme by her mother, who said that every other mode of treatment had been faithfully tried without success.

Her father was an utter disbeliever in magnetism, and the lady and family knew nothing more in regard to it than the common reports, and had little expectation of benefit from its employment ; they consented to the treatment as they would to any other harmless agency when the suffering was so intense, and the resources of medical art afforded no relief. Mr. Childs says : "I was informed that the young lady had for above two years been entirely out of health, suffering acute and capricious pains in the head, teeth and face, after the slightest exposure to cold, and pains at the pit of the stomach after every meal ; that about six months previously she had had an attack of brain fever ; that from defective capillary circulation she was always shivering from cold, and her countenance had assumed a perfectly exsanguineous aspect ; that she suffered so much from depression of spirit as to render life a burden ; and that for the last fourteen weeks she had been attacked every night by a pain extending over the right side of the head down to the shoulder, lasting for about five hours, and so acute as to almost deprive her of reason. Relief had been sought from change of air, diet, occupation and topical applications, and from alteratives ; yet up to the day in which I saw her, the disease had steadily persisted, and increased in violence.

At the first treatment by magnetism, the *physical effect* was most decided, and almost instantaneous. After the first sleep, the lady experienced a sense of warmth which she had not felt for months,—the first night, the agonizing pain, which had never once intermitted for fourteen weeks, wholly ceased, or was superseded by only a slight uneasiness, continuing but five minutes, instead of five hours. After two days this disappeared—and to the present time no sign of it has ever returned. The general improvement of her health is manifested by every indication of bodily health and vigor; at the end of six weeks the dull heavy languor of long disease had given place to the natural buoyancy of youth,—a colorless, lack-lustre complexion was replaced by the genuine hue of health,—the hair, which had nearly all fallen off, not only grew rapidly in length, but sprang up thickly over the whole head,—the appetite became perfectly good, and the digestion complete.

In fact the case presented a complete transformation from disease to health, and the whole character and appearance were so entirely changed, as to be scarcely recognized by her friends."

Her father, Dr. Larke, writes as follows .

"I am now most happy to state that her recovery is complete, that she enjoys perfect health and equanimity of spirit, and is everything I could wish her to be. This blessing I attribute to you, and I think I should be wanting in every proper feeling if I hesitated . to give my testimony to the case.

"BROOKE, Norfolk, March 13, 1845."

EPILEPSY.

Dr. Storer reports the two following cases in pp. 59, 60, *Mesmerism in Disease :*

"Was called to see Wm. Hodges, aged 19, a tailor of 10 South Moulton street. Three months previously he fell upon his left side on the ice, and, though his head was not struck, it was so shaken that he does not know how he got home. After reaching home he had repeated fits, decidedly epileptic, so that many men could scarcely restrain him ; and in four hours he was bled, relieving the fits for a day or two.

"But afterwards the least noise startled him, and caused a fit, even the sudden cough of a child. They were very frequent, and each left him weak for an hour. I found that there had been one fit on the 2d, the 8th, and the 14th, and two on the 15th. Mr. Wood mesmerized for me, treatment an hour daily, *except on Sundays.*

"He had no fit for a fortnight after commencement of the treatment. He was then frightened by a person behind him saying he would be run over ; 'felt ill—felt the fits on him,' but had no fit till evening, when, going up stairs in the dark, a cat jumped out and he was instantly seized with a paroxysm. It was very severe, stronger and longer than usual, and he tried to bite in it, felt ill and stupid all night, and did not sleep.

"He had no fit again till the *second Sunday* after this, and had another on the following Sunday. As they probably occurred from the omission of Mesmerism on Sundays, he was afterwards mesmerized every day till the middle of August, and he has never had a return to this hour, though very

often frightened enough to occasion them if there
was any predisposition left.

"He got married in August, 1842, and I saw him
in good health to-day (Dec. 24th)."

EPILEPTIC HYSTERIA, WITH LOCK-JAW AND CONTRAC-
TION OF ONE LEG.

"Maria Pearsey, twenty-five years old, a stout
and strong-looking person, was admitted under my
care into University College Hospital, June 25th,
on account of very frequent and violent fits, of an
epileptic and hysterical character.

"They occurred daily, and often many times in
the day; seized her suddenly, and produced perfect insensibility and very violent convulsions, so
that many could scarcely restrain her.

"Her right leg was firmly bent up nearly to the
body, and no force could bring it down.

"Her jaw was firmly locked; and we observed
that four of her front teeth had been drawn, and,
as we learnt, for the purpose of passing food into
her mouth to prevent starvation.

"This lamentable disease had commenced nine
years before, when she was sixteen years of age;
up to which time she had enjoyed perfect health.
She was then frightened by a young gentleman in
the house where she was a servant, jumping suddenly up before her, covered with a sheet, while
she was opening the cellar door. She fell insensi-

ble, and remained so for three days ; and from that
time had very violent fits, at first purely hysterical,
but at length more of an epileptic character.

" Not only had *private practitioners* failed to be of
any service to her, but she had been in vain four
months in St. George's Hospital, *four months* in
Guy's Hospital, *four months* in St. Thomas' Hospital, and *ten months* in Westminster Hospital, where
Mr. Guthrie gave her a *very large quantity of mer-
cury and salivated her severely*, and where Mr. White
was anxious to *cut off her bent up leg*.

" I determined to trust the treatment entirely to
Mesmerism ; she was treated half an hour daily.

" In *less than a fortnight* her jaw began to open ;
it opened now daily, and on July 12, *rather more
than a fortnight*, it opened widely.

" I have had two similar cases of lock-jaw of
some continuance in young women, since I culti-
vated Mesmerism ; and in both I succeeded perfectly. Had Mesmerism been used by those who
treated her previously, the poor girl might have still
been in possession of her upper and lower front
teeth. Not only, however, had the jaw opened
widely enough on the 25th for her to eat her dinner
like the other patients, but *her leg had relaxed so
much that the toes touched the ground*. Still she had
no power to move it, and it was in some degree
contracted.

" July 28th. *Her leg came quite down in the night.*
The fits lessened gradually, and *she went out well in
October ; and never had a return of her complaint.*

"No other sensible effect was produced than drowsiness, and sometimes a little sleep.

"If the passes were made quickly, she was distressed and a fit brought on.

"Being very nervous, and having suffered much, she was agitated and rendered hysterical and convulsed by any roughness of behavior, to which I regret she was often exposed.

"As the leg came down, splints and rollers were applied, to secure the ground gained ; and, in one of the other cases of lock-jaw to which I have alluded, I put a cork between the teeth as the jaw opened more and more, for the same purpose."

CASE OF ST. VITUS' DANCE.

Dr. Elliottson reports, through the columns of the *Zoist* for July, 1843, the case of Master Linnell of Northampton, nine years of age, which had baffled a great number of medical men, when at length application was made to Dr. Elliottson.

"On January 4th he was brought in a coach to me, and obliged to be carried into the house. Supported by his mother he walked with great difficulty from my dining room into my library.

"His debility was such that he could not stand a moment unsupported; his head hung on one side; his tongue out of his mouth, which constantly slobbered; his look was quite fatuitous; he could not articulate, making only inarticulate noises, and these with extreme difficulty ; even ' Yes ' or ' No,' were said in the strangest manner, so as to hardly be understood.

" He often fell into a passion at not being able to articulate; he ground his teeth, sighed greatly, continually blew bubbles of saliva from his mouth, and moved his tongue.

" The movements of the disease had lessened, so as not to be in proportion to his extreme muscular debility. He could use neither hand for any purpose, and scarcely ever raised the right. He was low-spirited and fretful, and often cried almost without a cause.

" His tongue was clean and moist, his appetite good, and his bowels in the most healthy condition; his pulse 74.

" He cried sadly at being brought to me, thinking I should give him loads of physic to swallow, and blister him, as others had done. I mesmerized him for half an hour by vertical passes before the face.

" He sat well supported in an easy chair, his head on his breast, *but he sat so quietly in comparison with his usual state that his mother noticed it.*

" He was treated in the same way daily until the 15th of February, when he was mesmerized for the last time."

Says the doctor in conclusion :

" Nothing could be more decisive of the power of Mesmerism than this case. The disease was getting worse and worse at the time I began. An effect was visible in a few days ; the benefit steadily increased, and from being a slobbering, idiot-looking child, his head hanging on one side, unable to speak, or stand unsupported, in three weeks he could stand easily, and walk five miles.

" Not a particle of medicine was given after the first day. The true gratitude of the boy and his mother was delightful."

DR. BRAID'S EXPERIENCE.

Mr. Braid, a celebrated surgeon of Manchester, England, to whom Dr. Carpenter refers in so complimentary a manner in his physiological works (who is also mentioned by Dr. Brown-Sequard in his lectures upon nervous force), devised a new method of exciting the primary magnetic condition, or the state of expectant attention. Mr. Braid styled his method " Hypnotism ;" and this appellation seems to suit some writers better than Magnetism or Mesmerism. Mr. Braid gives a history of a large number of cases treated by him, covering the senses of hearing sight and smell. Diseased conditions cured of tic-doloreux, spinal complaints, paralysis, rheumatism, both acute and chronic, nervous headache, epilepsy, &c.

We will quote a case or two under Mr. Braid's practice :

" Master J. B., thirteen years of age, was suddenly attacked with chilliness and pain all over his
body, on the evening of the 30th of March last.

" I was called to attend him the following day,
when I considered he had got a febrile attack from
cold, and prescribed accordingly.

" Next day, however, it had assumed a very different aspect. I now found I had got a severe case
of opisthotonos to deal with.

" The head and pelvis were rigidly drawn back,
the body forming an arch, and the greatest force
could not succeed in straightening it, or bring the
head forward.

" While the spasm never relaxed entirely, it frequently became much aggravated, when the head
was so much drawn back as to seriously impede
respiration. The legs were also sometimes fixed
spasmodically. The effect of the spasm in obstructing respiration and hurrying the circulation was
great, and seemed to place the patient in great jeopardy. The pulse was never less than 150, but during the paroxysm was considerably increased.

" It was evident I had got a most formidable case
to contend with, and that no time ought to be lost.
I therefore determined to try the power of Hypnotism, well knowing how generally such cases end
fatally under ordinary treatment.

" He was quite sensible, and the only difficulty
in getting him to comply with my instructions
arose from the recurrence of the severe spasmodic
attacks.

"In a very few minutes, however, I succeeded in reducing the spasm so that his head could be carried forward to the perpendicular, his breathing was relieved, his pulse considerably diminished, and I left him in a state of comparative comfort. In about two and a half hours after I visited him again, accompanied by my friend Dr. Cochrane.

"The spasms had recurred, but by no means with the same violence. Dr. Cochrane had no difficulty in recognizing the disease; but he did not believe any means could save such a case. He had never seen a patient hypnotized until that afternoon, and watched my experiment with much interest and attention. He seemed much and agreeably surprised by the extraordinary influence which an agency apparently so simple exerted over such a case. The pupil of the eye was speedily dilated, as if under the influence of belladonna; the muscular spasm relaxed, and in a few minutes he was calmly asleep." In conclusion Mr. Braid says: "I feel quite confident that without the aid of Hypnotism this patient would have died."

Another case of spasmodic affection—that of Miss Collins, of Newark. Her father writes:

"My daughter, sixteen years of age, had been afflicted for six months with a rigid contraction of the muscles of the left side of the neck to so great a degree that it would have been impossible to insert an ordinary card between the ear and shoulders, so close was their contact; and consequently she was rapidly being malformed. She had had the best advice to be procured in the country, and I had taken her to London with a written statement of the treatment previously employed, and had the opinion of Sir Benjamin Brodie, who approved of what had been done, but gave no hope of speedy relief.

"Hearing of Mr. Braid, surgeon of St. Peter's Square, Manchester, and with a letter written to that gentleman by Mr. Mayo, of London, I went with her, by the advice of Dr. Chawner, who, indeed, accompanied us, and placed her under the care of Mr. Braid on Thursday evening, the 24th of March. In less than a minute after that gentleman began to fix her attention, she was in a mesmeric (neurohypnotic) slumber, and in another minute was partially cataleptic. Mr. Braid then, without awaking her, and consequently without giving her any pain, placed her head upright, which I firmly believe could not, by any possibility, have been done five minutes before, without disruption of the muscles, or the infliction of some serious injury ; and, I am thankful to say, it not only continues straight, but she has perfect control over the muscles of the neck. A nervous motion of the head, to which she had been subject after her return from Manchester, has entirely ceased and she is at present in excellent health.

"At Dr. Chawner's suggestion, she was frequently watched while asleep, but not the slightest relaxation was observed in the contracted muscles."

(Signed)

JAMES COLLINS.

Mr. Braid adds : "After the lapse of a year, Mr. Collins was so kind as to write and inform me that his daughter continued in perfect health, with complete control of the muscles of the neck."

PAINLESS AMPUTATION OF THE THIGH, &C.

The Royal Medical and Chirurgical Society of London present an "account of a case of successful amputation of the thigh, during the magnetic state, without the knowledge of the patient, in the District Hospital of Wellow, Nottinghamshire."

The operator was W. Topham, of the Middle Temple : W. Squire, Ward Surgeon of Wellow Hall.

The patient was a laborer, six feet high, and forty-two years of age, named James Wombell. Full details were published at the time (November 22, 1842).

Dr. Storer reports the case of Mary Ann Lakin, of Leicester. She had been suffering from extensive disease of the knee-joint, which became so alarming as to render amputation necessary, even to prolong her existence.

The operation was performed by Mr. Tosswell, surgeon : the magnetizer was Mr. J. Collins.

She remained during the operation almost motionless, so that there was no occasion to support her on the table, as is usual on these occasions.

" On being put to bed after the operation, she was de-magnetized, and asked what she had felt. She said she had been dreaming that a person had tied something tightly round her SOUND leg, and that she felt angry, and gave him a kick, to make him desist. Her features did not change from the calmness first noted, and the witness, a surgeon, declares his belief that the girl, through magnetism, avoided the horrors of the operation."

It would be impossible to give a synopsis in this brief work of the hundreds of cases of painless surgery recorded by even one practitioner, Dr Esdaile, in India, whose interesting work on Mesmerism has already been alluded to.

One or two will exhibit sufficiently the complete character of the anæsthetic effects.

" JUNE 3d.—Teencowrie Paulit, a peasant, aged forty, was admitted to the hospital. Two years ago, he began to suffer from a tumor in the *antrum maxillare;* the tumor had pushed up the orbit of the eye, filled up the nose, passed into the throat, and caused an enlargement of the neck.

" I was very desirous to reduce him to a state of insensibility before operating upon him, and for the last fortnight my assistants have all perseveringly tried it, but without inducing sleep even. Indeed, from the tumor obstructing his throat, he has hardly slept for five months. Having ascertained that he was easier when sitting, I took him in hand myself, to-day, and entranced him in a chair. In half an hour the man was catalepsed, and in a quarter

more I performed one of the most severe and protracted operations in surgery; the man was totally
unconscious.

"I put a long knife in at the corner of the mouth,
and brought the point out over the cheek bone dividing the parts between; from this I pushed it
through the skin at the inner corner of the eye, and
dissected the cheek back to the nose.

"The pressure of the tumor had caused the absorption of the anterior wall of the *antrum*, and on
pressing my fingers between it and to the bones,
it burst, and a shocking gush of blood and brain-
like matter followed.

"The tumor extended as far as my fingers could
reach under the orbit and cheek-bone, and passed
into the gullet—having destroyed the bones and
partition of the nose. No one touched the man,
and I turned his head into any position I desired,
without resistance, and there it remained till I
wished to move it again; when the blood accumulated, I bent his head forward, and it ran from his
mouth as from a spout. The man never moved nor
showed signs of life, except an occasional indis-
tinct moan; but when I threw back his head, and
passed my fingers into his throat to detach the
mass in that direction, the stream of blood was
directed into his windpipe, and some instinctive
effort was necessary for existence; he therefore
coughed, and leaned forward to get rid of the
blood; and I had supposed he then awoke.

"The operation was by this time finished, and he

was laid on the floor to have his face sewed up, and while this was doing, he for the first time opened his eyes.

"JUNE 4th.—The man declares, by the most emphatic pantomime, that he felt no pain while in the chair, and that when he awoke, I was engaged in sewing up his face, on the floor ; so that the coughing and forward movement to get rid of the blood, were involuntary, instinctive efforts to prevent suffocation.

"JUNE 6th.—The dressings were undone to-day, and the whole extent of the wounds in the face has united completely by the first intention. He is out of all danger, and can speak plainly ; he declares most positively that he knew nothing that had been done to him till he awoke on the floor, and found me sewing up his cheek."

Here is a statement from himself, translated from Bengalee :

"For two years I have suffered under this disease, and scarcely slept for five months.

"On the 19th of May, I came to the Imambarah Hospital, and three or four persons tried to make me sleep, but all in vain. On the 3d of June, Dr. Esdaile having kindly undertaken my case, with a great deal of labor, made me sleep, and took something out of my cheek, which at that time I did not perceive.

"After the operation I did not sleep for two nights, but after the third day I have slept as usual.

"TEENCOWRIE PAULIT,

of Madra."

If patients are fortunate enough to sleep some time after the operation, they not only feel no pain on waking, but none subsequently even.

Dr. Esdaile records the removal of a large number of tumors in addition to the above, of which one more specimen will perhaps suffice.

"OCT. 25th, Gooroochuan Shah, a shop-keeper, aged 40. He has a 'monster tumor' which prevents him from moving; its great weight, and his having used it for a writing desk for many years, has pressed it into its present shape. His pulse is weak, and his feet œdematous, which will make it very hazardous to attempt its removal; but with such an appendage life is literally a burden. He became insensible on the fourth day of mesmerzing, and was drawn with the mattress to the end of the bed (my usual mode of proceeding); two men then held up the tumor in a sheet, pulling it forward at the same time, and, in the presence of Mr. Bennett, I removed it by a circular incision, expedition being his only safety. The rush of venous blood was great, but fortunately soon arrested; and after tying the last vessel, the mattress was again pulled back upon the bed with him upon it, and at this moment he awoke.

"The loss of blood had been so great that he immediately fell into a fainting state, and it took a good while to revive him. On recovering he said that he awoke while the mattress was being pulled back, and that nothing had disturbed him. The tumor weighed EIGHTY pounds, and is probably the largest ever removed from the human body. I think it extremely likely that if the circulation had been hurried by pain and struggling, or if the shock to the system had been increased by bodily and mental anguish, the man would have bled to death, or never have rallied from the effects of the operation.

" But, the sudden loss of blood was all he had to contend against; and though in so weak a condition, he has surmounted this, and gone on very well.

" DEC. 1st.—Has been allowed to go home at his own request ; the wound is filling up slowly, for want of integument."

No single man has done so much, practically, in Magnetic-Surgery as Dr. Esdaile, and as his practice is so often quoted by contemporary and later writers, we would again urge our readers to examine his very valuable work. His chief practice was surgery ; but he also gives many other cures by magnetic treatment, among which are enumerated Nervous Headache, Tic-douloureux, Spasmodic Colic, Acute Inflammation of the Eye, Chronic ditto, Convulsions, Rheumatism, Lumbago, Sciatica, Pain in the Crural Nerve, Palsy of Limbs, ditto of half of the body, and many other forms of disease cured by this agency. He says, further, as a result of his experience :

"In chronic inflammation it is a useful discutient, gently stimulating the nerves and capillary vessels of the part to more healthy action ; and for this purpose local magnetism is only required.

"The chronic exhibition of magnetism as a general tonic in diseases of debility promises to be of great service, especially in functional derangement of the nervous system, and I am hopeful that we have at last got a DIRECT nervous remedy, hitherto a 'desideratum,' in medicine.

"In palsy from weakness of the nerves, it promises to assist us greatly, and Dr. Elliottson has recorded many cures of cases he could not have managed before. All who venture to confess the truth to themselves know how miserably impotent for the cure of palsy and nervous diseases generally are the medical means hitherto employed. If we succeed, we often cannot tell why, and the connection of cause and effect is very uncertain.

"But in the chronic treatment of palsy, by *magnetism alone*, the patient often feels and shows early and continued improvement under the action of this natural remedy ; and we cannot refuse to believe that it is the exciting cause ; we must believe that it is, or own that it is a spontaneous cure of a commonly incurable disease ! But both the practitioner and patient must remember, that Mesmerism is no exception to the general rule, that a chronic disease requires a chronic cure. Much patience and labor will be required in the mesmeric treatment of paralytic affections."

So wrote this honest, frank, outspoken physician. His powerful statement of thoroughly attested facts thrilled all Europe, and this painless surgery which he practiced and endorsed would to-day be the reliance of the profession, were it not for the almost contemporaneous introduction of ether and chloroform, which surgeons find more convenient to administer, preferring the *modicum* of risk which attaches to it, to the delays which magnetism would involve.

But, assuming that convenience justifies this position, there are, after all, some lives that hang upon threads so slender, that it is positively impossible to administer either of the above-named drugs, where magnetism might be used with perfect confidence, and an operation otherwise impossible could be safely attempted. False professional pride, recklessness, ignorance or stupidity will alone prevent the frequent use of an agent so effectual, and proved and admitted so SAFE, as vital magnetism.

ENGLISH PRACTITIONERS AND ADVOCATES.

As early as 1848 Mr. Sandby, in his interesting work upon the subject, says : " There is hardly a county in England where it is not now practiced. From York to the Isle of Wight—from Dover to Plymouth, there can be produced a chain of evidence and a list of cures. * * *

" Mr. Atkinson has been eminently successful in his treatment of numerous diseases, some of which generally defy all human skill, among which were three cases of that fearful malady, TIC-DOULOUREUX;

one of which was of *ten years' standing*, and the other
two of several years' duration. He has also cured
many cases of fits, of hysteria and want of sleep,
and of those determined nervous and sick head-
aches, which seldom yield to remedial action; he
has been equally successful in various acute nervous pains, and contractions of limbs, asthma, fever,
long standing cough, affections of the heart and
spine, injured sight, deafness, melancholia, rheumatism, tooth-ache, indigestion, and functional obstructions."   Mr. Atkinson was one of the best
early practitioners.

Mr. Chenevix, a well-known name in the literary
world, and a man of scientific training, was also a
very successful operator, and at the time of his sudden death by acute disease, had in preparation a
work demonstrating his own experience upon some
442 patients. He says: "There was hardly one
instance where disease existed, that relief was not
procured."

" He reports the cure of a case of epilepsy and
spasmodic pain of six years' standing in twenty-
one treatments. He succeeded in three other cases
of a similar character completely, and procured
immense relief in eight others. He also cured
seven cases of worms, and was of great service to
several inmates of the Wakefield Lunatic Asylum."

Colonel Sir Thomas Wiltshire, commanding at
Chatham, was also an eminent practitioner.

Earl Stanhope was also a practitioner, and a gentleman of great Christian kindness and philan-

thropy. He notes many cases of cure in his own experience.

Capt. Valiant, of Chatham, was a very powerful and successful operator, as was also Mr. Baldock, who found it very successful in palpitations of the heart and severe headaches; and amongst other cures he reports that of Robert Flood, of Caister in Lincolnshire, who suffered for several years from disease in one of his kidneys.

He had been under the care of several eminent physicians, and was finally admitted to the hospital in London; his pains were so acute that he could not leave his bed until the day was far advanced, and even then he was compelled to recline several times before retiring for the night. Mr. Baldock restored him to health, by magnetic treatment, in three months.

Mr. Majendie, of Hedingham Castle, Essex, was equally prominent and successful.

Mr. Topham of the Temple was also a very prominent operator, and was the magnetizer in the case of Wombell, whose thigh was painlessly amputated; and he was very successful in the treatment of Epileptic fits.

Mr. Thompson, of Fairfield House, near York, was a famous operator, and gives a great many astonishing cures.

He mentions a child nine years of age, who had a diseased knee of a scrofulous character, from which he suffered "intense agony, was unable to rest day or night, with loss of appetite, great inflammation extending above the knee, the knee enormously enlarged, with extensive suppuration on the inside of the knee, a high state of fever at the time, with a hectic flush upon his cheeks, with quickness of breathing, and a short cough." By local magnetizing the symptoms were ameliorated in half an hour, the child became still and calm, and smilingly described a "heat coming out of Mr. Thompson's fingers," which had taken the pain away. The child rapidly improved, absorption of the matter took place, and in a month's time he was able to use the limb; sleep was not induced in this case.

In another case treated by Mr. Thompson, a gentleman had been suffering for nine consecutive days from severe rheumatic fever, with acute pain in the shoulders, arms, hands, loins, legs and knees; excessive fever, profuse night-sweats, caused by the agony of pain.

The following is the patient's description of the magnetic treatment:

"In less than twenty minutes you had nearly CHARMED away all the pain and restored warmth and feeling to my feet. You then put me to sleep; the delightful sensation of that sleep, after such extreme pain, I can scarcely describe. When you awakened me, I felt like another person. The fever was reduced, and the pain was gone. In four days I was down stairs. Every time you mesmerized me I felt as it were a new life (pp. 151 Sandby's Mes.)."

Captain Anderson, of the Royal Marines, a resident of Chelmsford, another powerful magnetizer, reported many cases:—

One of Mrs. Raymond, residing at Chelmsford, who had suffered for nine years from spinal complaint, being confined to her sofa, and unable to be moved day and night; she had also lost the use of her voice. Her sufferings were dreadful. Blisters, caustic plasters, leeches, setons, and medication endless had been tried with no substantial benefit.

At first this afflicted lady laughed at the idea of being relieved by such apparently inadequate means ; but having finally consented to give it a trial at the hands of Captain Anderson, she states her experience in these words :

"I am now able to walk out daily, alone and unassisted. I am regaining my speech, and I am free from pain, sleep soundly, and take no medicine, and am now seldom mesmerized." (For full particulars of this case, see the *Zoist*, vol. 2, p. 82.)

Dr. Engledeu, of Southsea, practiced it extensively in his profession, and has been a very warm advocate of its merits.

Mr. Weeks, a surgeon at Sandwich, devoted a large portion of his time to the judicious application of magnetism, and says, after two years' practice, that "its use has been demonstrated after the usual modes of treatment, and in some instances abundance of quackery to boot, had utterly failed, and rendered the case more inveterate and distressing." Among the cases he mentions, are some of

dyspepsia, habitual constipation, paralysis, slug-
gish condition of the hepatic system, hypochondriasis, muscular contractions, stubborn and otherwise hopeless cases of rheumatism, local pains, and
severe forms of neuralgia, cases of general languor
and debility, without manifest cause ; as also of
deafness, and painless surgery, besides several affections of an anomalous character. '

Mr. Weeks has ranked high, with an immense
practice. Mr. Prideaux, of Southampton, is another great practical magnetist, and reports several
cases of St. Vitus' dance, treated in his practice by
this agency. He is a medical man, and bears high
testimony in its favor.

Mr. Janson, of Pennsylvania Park, President of
the Exeter Literary Society, a gentleman of high
literary attainments, bears valuable testimony, and
has met with marked success in the treatment of
tic-douloureux.

Mr. Holm, of Highgate, a gentleman of philan-
thropic character, says that he has found it a very
efficacious remedy in epilepsy, rheumatism, brain
fever, head-aches, and many neuralgic disorders.

Mr. Charles Childs, of Bungay, surgeon, says :
'' I have practiced Mesmerism about four years ; in
this period I have proved its *unquestionably beneficial
results in several of the most afflictive maladies.*'' Several of his cases are published, including many in
painless surgery. (See *Zoist*, p. 36, vol. 3.) A large
number of cases are also credited in the same journal
to Mr. Chandler, surgeon, of Rotherhithe, amongst

which is a rapid and perfect cure of a case of insanity, when all other remedies had failed.

Mr. Purland, of Mortimer street, Cavendish Square, has practiced the art with great success and reports a large variety of cases in his practice, amongst which are asthma, hysteria, lameness, deafness, &c.

Mr. Boyton, Surgeon of Wattington, in Oxfordshire, is a gentleman of skill and reputation in the profession; after describing many cures, he says : *" It strengthens* the nervous system, improves the digestion, and tranquilizes the mind."

Dr. Wilson, of the Middlesex Hospital, well known through his publications on the subject, reports a case of insanity cured by Magnetism.

Dr. Ashburner, of Grosvenor street, London, late physician to the Middlesex Hospital, Dr. Storer, of Bristol, Dr. Simpson of York, Dr. Arnott of Edinburgh ,Mr. Johnston, surgeon, 22 Saville Rowe, Sweden, have published remarkable cases.

Mr. Newnham, surgeon, of Farnham, the author of " Human Magnetism," was requested to write a paper *against* the practice of Magnetism, and was furnished with materials for the purpose; but investigation soon convinced him of the value of the power, and he became one of its stanchest advocates.

Mr. Sandby refers to Symes of Grosvenor street, London; J. Hands, Duke street; Decimus Hands, Thayer street; Morgan, Bedford Row; Flinthoff, Great Tichfield street; Clarke, Kingsland Road;

Case, of Fareham; Adams, of Lymington; Choter, of Norwich; Weddell of Scarborough; Nixon, of Wigton; J. B. Parker, of Exeter; Sargeant, of Reigate, Surrey; Luxmore, of Alphington, Exeter; Hollings, of Leicester; Vivian, of Woodfield, Torquay; Briggs, of Nottingham Place; Mulholland, of Walsall, who reduced a wen of eleven years standing, the size of a goose egg, so completely, that it requires astute observation to detect it, Stenson, of Northampton; Summers, of Chatham, who acted successfully upon a case of obstinate hernia, by Magnetism; Brindley, of Stourbridge; Tubbs, of Upwell Isle, in Cambridgeshire, with a very large practice and a long list of remarkable cures; Dr. Owens, of Stourbridge; with a host of non-professional advocates.

"Here then," says Mr. Sandby, "is a train of witnesses consisting of men of ability, and honorable standing, from whose reports the facts of magnetism might be confidently predicated, and this list might be swelled to any extent.

"What an amount have we here of happiness conferred! What a mass of pain, of sickness, of sorrow, lightened or removed! Here at length are a few pleasing pages in the long sad chapter of human life! Here, at last, is a delightful study for the philanthropist and the Christian."

To a far greater extent than in Great Britain has its application been made on the continent; and Dr. Georget, the celebrated French physiologist, justly remarks of the character of its advocates :

"It is a very astonishing thing that animal magnetism is not even known by name among the ignorant classes ; *it is among the enlightened ranks that it
finds support, learned men, naturalists, physicians, and
philosophers.*"

PREJUDICE, IGNORANCE AND CHARLATANRY NO LON-
GER PERMISSIBLE.

The mass of evidence by which this practice
stands fortified to day, has not been accumulated,
however, without hard blows being exchanged between its advocates and opponents ; the battle
waged furiously for more than thirteen years, but
there could be but one issue between fact and ignorant opposition ; and simple prejudice ultimately
was compelled to give way before the crushing
array of indisputable evidence in favor of magnetic
treatment. In a vast variety of diseased conditions, especially in the field of nervous disorders
its results were most astonishing.

Many men were hunted as wild beasts for acting
up to their convictions ; notably, Dr. Elliottson,
who, in consequence of the narrow-minded policy
pursued by other members of the faculty, was induced to withdraw his name from the Professor-
ship of the University of London, because of their
opposition to the practice in that institution. The
patients themselves were willing recipients, and
the most signal benefits were being daily experi-

enced, and the Academies of Paris and Berlin had
not thought the question beneath their notice.

This outrage, however, was not accomplished
without the strenuous opposition of four members
of the council, comprising the names of Lord
Brougham, Sir L. Goldsmith, Mr. Tooke, and Mr.
Bishop, who placed too high a value upon the standing and character of Dr. Elliottson to let this ungenerous action pass without a protest. The experience of Dr. Elliotson was duplicated in many
other directions ; but these frantic efforts in opposition carried their own defeat, in the final contest.
Hospitals and infirmaries had to be established in
the earlier days of investigation for this special practice until the wider diffusion of information caused
its merits to be admitted, and the general practice
won its way in public estimation to such a degree,
that this necessity passed away.

The best physiologists of Europe have long since
acknowledged the claims of this branch of therapeutics, and to-day can calmly state its merits without the risk of personal slanderous attacks, which
is of itself a great gain upon the past, and has
made its extensive application abroad an easy matter, so that its standing is well assured. / But how
stands the practice in our own country ? While
Europe has been ablaze with investigation, we have
done little ; we have thus escaped much of the bitterness of discussion, and may reap all the advantages of the practice as the result of other men's
labors, still, we have lacked the stimulus of this

public agitation, and exhibit a supineness and want of enterprise truly amazing, in view of the transcendent importance of a subject so vital to the welfare of humanity.

It is true that it has been utilized quite extensively in this country; but amongst medical men there has been an almost total lack of information upon the subject, to characterize it by no stronger term, that is very extraordinary.

There are, doubtless, reasons for this, so far as individual action is concerned. In the first place, our medical universities have almost totally ignored the subject, mainly, it is believed, because of the supposed difficulty of carrying it into general practice. This has prevented inquiry for its literature, which, while so abundant abroad, is scarcely to be procured here. France and Germany alone have put forth over fourteen hundred publications on the subject; but it is almost impossible to find a single volume upon any bookseller's shelves in our land treating of vital magnetism.

Physicians do not hesitate to express their ignorance in regard to it, while they may have a general impression of its benefits.

But, perhaps the most potent cause of this lack of information here has arisen from the prejudice awakened by the lecturers who, many years since, presented upon public platforms, in various forms of buffoonery, the psychological phenomena so readily induced in a certain class of *magnetic sensitives.* These merely farcical exhibitions have been proclaimed as the usual effects of magnetism, and have been so accepted by the public ; and the public being so misinstructed, have scarcely dreamed of the serious side of the question, nor of the vast interests which were involved in it.

While such entertainments may not have produced physical injury to any one, they have done immense mischief by bringing into ridicule what should have been guarded most carefully from reproach. The same adverse influences existed, and were felt in England, but their effects were largely neutralized by the able advocates of the higher and grander aspects of this noble science.

Says Harriet Martineau, in one of her letters : "I believe there is no doubt that the greatest of all injuries done to Mesmerism is by its itinerant advocates.

" This appears to be admitted by everybody but the itinerants themselves ; and none lament the practice so deeply as the higher order of Mesmerists."

It is true the blame of this desecration rests with the learned men, who ought to have shown themselves wise in relation to a matter so serious, and to have taken the investigation into their own hands. It is they who are answerable for having turned over the subject to the fanatical and the vulgar. It is they who have cast this jewel of knowledge and power into the lap of the ignorant ; and no one can wonder that it is bartered for money and notoriety."

Every able writer of those times struck heavy blows at the men and their practices, who simply made a *show* of the mental phenomena, and were incompetent even to state the higher uses of this force, and thus to a great extent the mischief was prevented in England which lamentably followed in this country, for lack of strong friends to intelligently stand for the truth.

We are gradually emerging, however, into a better light, and the confident utterances of our best physiologists are awakening a spirit of inquiry among physicians in medical conventions, and in the public mind, which promises valuable results in the immediate future.

XII.

THEORIZING.

As I said, my object is not to set up a *theory* in regard to the origin or causation of magnetic phenomena, a rock upon which most writers have split.

The world has been deluged with theories ; and if my readers would pursue a phantom I will place

them in possession of sources of INFORMATION ; but
this is not the *rôle* which I have prescribed for myself.

Vital Magnetism is a science based eminently on
facts, not on opinions ; and these are as uniform
and as well authenticated as facts in physiology,
or the phenomena of Electricity, Light, Heat, or
Mineral and Terrestrial Magnetism. On these *facts*
it rests its claims for consideration, adoption, and
general utilization.

\ As Miss Martineau has well said, in pleading for
its consideration, when inconsiderate friends demanded a theory from her :

" ' Tell me what Mesmerism is, first, and then I
will attend to it,' has been said to me, and has been
said to many others, who, declaring Mesmerism to
be true, have no theory as to its nature—no con-
jecture as to the scope of its operations.

" Some ask this in ignorance, others as an eva-
sion. Wise inquirers will not ask it at all till a
vast preparatory work has been achieved, which it
is both unphilosophical and immoral to neglect.
There are hospitals among us, where it may be ascertained whether insensibility to pain can be produced.

" There are sufferers in every one's neighborhood
whose capability of recovery by Mesmerism may
be tested.

" Let experience, carefully obtained, be wisely
collected and philosophically communicated.

" If found untrue, Mesmerism may then be ' exploded,' which it can never be by mere ignorant scorn and levity.

"If true, the world will be so much the better. When we consider that no physician in Europe above forty years of age, when Harvey lived, believed in the circulation of the blood, we shall not look for any philosophical inquiry into Mesmerism from established members of the profession, whose business it is to attend to it ; but happily, the young never fail. There always is a new generation rising up to emancipate the world from the prejudices of the last (while originating new ones), and there are always a few disinterested, intrepid, contemplative spirits, cultivating the calm wisdom and bringing up the established convictions of the olden time, as material for the enthusiasm of the new, who may be relied on for maintaining the truth till they joyfully find that it has become too expansive for their keeping. The truth in question is safe, whether it be called Mesmerism, or by another and better name.

"Everybody interested in a great discovery is under a strong temptation to theorize too soon; and those who oppose, or do not understand Mesmerism, are forever trying to get us to theorize prematurely. From the first day that my experiment was divulged, to the present, the attempt has been renewed till the application to me to announce a theory has become so ludicrously common, that I am in no danger of falling into the trap. I have had not only to refuse to propose even a hypothesis, but to guard my language so carefully, as that by no pretense of an inference could any be ascribed to me.

"I could wish that all who, like myself, knew personally but a few facts (however clear), were as careful about this as the occasion requires. Their notions of a transmission of a fluid, electric or other —of a conditional excitement in human beings, of a power of control or stimulus of their own vital functions—of the mesmeric power residing in the will of the mesmerist, or in the imagination or will of the patient; of some sympathetic function, express but obscure, and assigned to some unexplored region of the brain—these notions and many more, may each suit the phenomena which have come under the notice of the expounders; but no one of them will hold good with all the facts that are established.

"The phenomena are so various, that it seems to me most improbable that we can yet be near the true theory, to say nothing of what is very obvious—that the suppositions offered are little but words."

These remarks of Miss Martineau apply with equal force to-day, for we are no nearer the solution of the problem of causation in regard to this *human force* than we are of the many other great natural forces by which we are surrounded, and which men have almost ceased to speculate about.

For the same reasons I have entirely ignored in this volume the so-called "higher class" of mental

or psychological phenomena induced often-times
in magnetic sensitives ; these matters are very interesting, but I have failed to see the practical relation between somnambulism, clairvoyance, intro-
vision, pre-vision, ecstasy, phreno-Mesmerism, mind
reading, &c., and the use of magnetism as a healing
remedy. It is evident that the discussion of these
uncertain and variable phenomena by most of the
writers on vital magnetism, has bewildered, con-
fused and oftentimes disgusted seekers after the
true practical points involved in the application of
magnetism in the treatment of disease. Even physicians have been led off from the real practical issue by these exciting and bewildering questions,
and in giving the results of their investigations,
they have mixed up moral, spiritual and religious
speculations with their statements of facts in relation to the cure of disease, to such an extent that
their value has been diminished, and often destroyed.

The mental phenomena of magnetism should, in
my judgment, receive separate investigation and
statement, and the facts relating thereto should be
arranged by themselves. My objects are :

First, to call the attention of physicians to this
agency for the treatment of many otherwise incurable forms of disease, and to stimulate investigation, and,

Second, to show anxious sufferers that there is
the possibility of a happy release both from drugs
and pain.

XIII.

There still remains much of misapprehension in the average mind, in regard to the supposed scope and power of magnetism ; some affect disbelief, while others admit the facts, but timidly apprehend dangers, moral, spiritual and physical, which do not exist, or at least not to the extent that weak and superstitious persons suppose.

There being no public attacks in our day against the practice and use of vital magnetism as a curative power, we do not hear, except in a casual way, of these foolish fears and apprehensions ; but they do exist ; and to meet these objections, the arguments being precisely the same as those used thirty or forty years ago, let us see how Miss Martineau met the objections, for her opinions and experiences were based upon personal knowledge, and never lose their freshness or application.

She says : " I have less sympathy with those who throw down their facts before the world, and then despise all those who will not be in haste to take them up than with some I know of, who would seriously rather suffer to any extent, than have recourse to relief, which they believed unauthorized ;

who would rather that a mystery remained sacred than have it divulged for their own benefit; who tell me to my face that they would rather see me sent back to my couch of pain than witness any tampering with the hidden things of Providence.

"There is a sublime rectitude of sentiment here, which commands and wins one's reverence and sympathy; and if the facts of the history and condition of Mesmerism could bear out the sentiment, no one would more cordially respond to it than I— no one would have been more scrupulous about procuring recovery by such means—no one would have recoiled with more fear and disgust from the work of making known what I have experienced and learned.

"But I am persuaded that a knowledge of existing facts clears up the duty of the case, so as to prove that the sentiment must, while preserving all its veneration and tenderness, take a new direction, for the honor of God and the safety of man.

"As for the frequent objection brought against inquiry into Mesmerism, that there should be no countenance of an influence which gives human beings such power over one another, I really think a moment's reflection, and a very slight knowledge of Mesmerism, would supply both the answers which the objection requires. The power is abroad, and ought to be guided and controlled.

"Next, this is but one addition to the powers we have over one another already; and a far more slow and difficult one than are many which are

safely enough possessed.   Every apothecary's shop
is full of deadly drugs ; every workshop is full of
deadly weapons ; wherever we go, there are plenty
of people who could knock us down, rob and mur-
der us ; wherever we live, there are plenty of people who could defame and ruin us.

"Why do they not? Because moral considerations deter them.

" Then bring the same moral considerations to
bear on the subject of Mesmerism.

" It is a mistake to suppose that Mesmerism can be
used at will to strike down victims, helpless and
unconscious, as laudanum does, except in cases of
excessive susceptibility from disease ; cases which
are of course under proper ward. The concurrence
of two parties is needful in the first place, which is
not the case in the administration of narcotics ;
and then the practice is very uncertain in its results
in most single occasions ; and again, in the majority of instances, it appears that the intellectual and
moral powers are more, and not less, vigorous than
in the ordinary state.

"As far as I have any means of judging, the
highest faculties are seen in their utmost perfection
during the mesmeric sleep; the innocent are stronger in their rectitude than ever, rebuking levity,
reproving falsehood and flattery, and indignantly
refusing to tell secrets, or to say or do anything
they ought not. The volitions of the mesmerist
may actuate the movements of the patient's limbs,
and suggest the material of his ideas, but they

seem unable to touch his MORALE. In this state the MORALE appears supreme, as it is rarely found in the ordinary condition.

"And now one word of respectful and sympathizing accost to those reverent and humble spirits who painfully question men's right to exercise faculties whose scope is a new region of insight and foresight. They ask whether to use these faculties be not to encroach on holy ground, to trespass on the precincts of the future and higher life. May I inquire of these, in reply, what they conceive to be the divinely appointed boundary of our knowledge and our powers? Can they establish, or indicate, any other boundary than the limit of the knowledge and powers themselves? Has not the attempt to do so failed from age to age?

"Is it not the most remarkable feature of the progress of Time that, in handing over the future unto the past, he transmutes its material, incessantly and without pause, converting what truth was mysterious, fearful, impious to glance at, into that which is safe, beautiful and beneficent to contemplate and use; a clearly consecrated gift from the Father of all to the children who seek the light of his countenance?

"Where is his pleasure to be ascertained but in the ascertainment of what he gives and permits, in the proof and verification of what powers he has bestowed on us, and what knowledge he has placed within our reach? While regarding with shame all pride of intellect, and with fear the presumption of ignorance, I deeply feel that the truest humility is evinced by those who most simply accept and use the talents placed in their hands; and that the most childlike dependence upon the Creator appear in those who most fearlessly apply the knowledge he discloses to the furtherance of that great consecrated object, the welfare of the family of man.''

These moderate, tolerant and intelligent views which I have quoted from this gifted woman's writings, were published thirty years ago in the *Atheneum*, and are presented with a peculiar charm and forcible directness.

Thirty years of experience confirmed the truth of these utterances, and failed to develop the dangers which superstitious objectors anticipated and predicted.

Among the most graceful writers on the subject is the Rev. Chauncy Hare Townshend of Trinity Hall, Cambridge, England, a very scholarly gentleman, who published in 1844 his work entitled *Facts in Mesmerism*, heretofore quoted from. Mr. Townshend's researches were mainly in the realm of mental phenomena, and upon this branch of the subject he will be found as clear and interesting as any other writer.

Among other gentlemen of scientific attainments who certified to Mr. Townshend's *facts* and power, was Professor Agassiz, who was himself magnetized by Mr. Townshend.

But, while this able writer and investigator devotes his main efforts to developing psychological phenomena, he saw its practical effects, especially on the continent, and fully and freely endorsed its strongest claims in this direction. He presents a beautiful view of its higher moral aspects worth quoting.

His argument will be found in the " notice " to the second edition of his " Facts."

" The metaphysical deductions which may be . drawn from mesmeric phenomena, have always appeared to me, in themselves, to constitute a sufficient and a noble answer to the inquiry: In what can Mesmerism contribute to the welfare or knowledge of man ?

" The light which this agency throws upon the mysteries of our nature, upon some of the profoundest truths of religion ; the indications which it alone presents of a possible separation between soul and body, and of the predominating powers of the former—must ever vindicate the sincere inquirer into its laws from the charge of trifling with an idle subject; granting even that he should confine himself to half his theme, and decline touching upon the relief, which, in a medical point of view, Mesmerism is capable of affording to the evils of suffering humanity.

" But it is the peculiar happiness of Mesmerism not to be forced to rely upon any one solitary and partial claim to notice and consideration.

" Its roots are cast deeply and extensively into the general ground of humanity.

"Where the metaphysician leaves it, the man of science may take it up; and when Science has gathered in its store of valuable facts, illustrative of all her noblest theories, it can still afford an ample harvest to him who would practically ameliorate the condition of his fellow-beings.

"If it fail to win the heart of the materialist from his barren creed by its revelations of Mind, as the only real source of motion—as all, indeed, that can be named—Power;—if vainly it preach to the sceptical or the careless, of Truth, Righteousness, and Judgment to come—it ought, at least, to recommend itself to every friend of man by the palpable benefits it is calculated to bestow in cases of disease, which are either unaffected by, or opposed to, the action of other remedies.

"That this pretension, as we will modestly call it, on the part of Mesmerism, to rank at least as a means of cure, should not universally prevail to win for it a patient and unbiased trial and examination, is indeed a proof of the force of prejudice to turn us even from our habitual ways of proceeding. I can well understand that Mesmerism, reasoning of things beyond the tomb, should meet with contumely or neglect ; for we seem, indeed, to be so constituted that the present and the physical lord it over the future and the intangible ; so that a power, professing to heal diseases, will interest men generally more than the most striking revelations regarding their moral destiny ; and it may be reasonably doubted whether our Redeemer's Sermon on the Mount made anything like the impression which the healing of the leper, at its foot, produced upon the assembled multitudes. But, that Mesmerism, proclaiming—" I can cure !"—should be coldly or partially received, does, I own, move me to wonder.

"Ordinarily, all that holds out even a hope of relief to any ailment whatever, whether real or fancied, is welcomed with a blind rage and stupid credulity. Every new name of every old quack remedy is certain to evoke a crowd of self-devoted persons, who are most ready to try on their own poor bodies the efficacy of the nostrum. In the very teeth of verdicts that denounce the drug as murderous, the fond disciples of a favorite pill will yet continue to obey the voice of the inventor, that says : ' The more you take the better ' (*id est* : ' for my own pocket ') and will greedily swallow the favorite boluses by fifties, and by hundreds.

"In the case of Mesmerism alone, the most startling benefits, the most alluring promises, on the one side—persecution even, and the anathemas of the diplomatized on the other—have failed to stimulate the public mind into anything like desire or enthusiasm. Perhaps even so, it is well. All slowly acquired possessions are the surest, and that, which owes nothing to the breath of popular applause, may consider itself independent of popular fickleness. The part that now remains to the friends of Mesmerism is, by truth and sobriety, to win their way into the territory of public belief, where every step is grudgingly accorded, and gradually to persuade men to their own benefit by setting before them unceasing examples of the curative powers of the distrusted agency. It is in this sphere of its operation especially, that my experience has become enlarged ; and as I have elsewhere sufficiently insisted on the metaphysical importance of Mesmerism, I would here confine myself to stating a few facts relative to its remedial energy, which have either fallen under my own observation or have reached me through unsuspected sources. I have watched the effects of mesmeric treatment upon a suffering friend, who was dying of that most fearful disorder—Lumbar Abscess.

" Unfortunately, through various hindrances, Mesmerism was not resorted to till late in the progress of the disease, so that, of course, that it should effect a cure was out of the question. It should ever be held in remembrance that Mesmerism does not profess to work miracles.

"It cannot restore a decayed bone to its integrity, or re-create a missing part ; but it can benefit, even where it cannot save.

" And how much is it to say of a power—that it is remedial even where not curative, and that in cases where it fails to re-kindle life, it can smooth the passage to the grave, and mitigate the horrors of physical pain ! This I have gratefully witnessed in the case of my suffering friend. He was at the time of the worst part of his illness, residing near Inspruck, and was fortunate enough to command the mesmeric services of Dr. Ennemoser, the author of several high and useful works on Mesmerism; a man whose experience in this science is fully equaled (as it ought to be in medical men) by his knowledge of therapeutics in general.

" I have no hesitation in saying, that, under God, the life of my friend, R. T., was prolonged, at least, two months by the action of Mesmerism. The improvement in all his symptoms, from the date of the first experiment, was too marked to be mistaken, or to be considered as the result of other causes.

" At the time the mesmerising was undertaken, the patient was in a state of continuous fever, totally without appetite, and suffering almost constant pain from thirteen wounds, which were partly the result of his disorder, partly of abrasion from his being nearly always in a reclining position, either in bed or on a sofa.

" His thirst was great, his tongue white and dry, and his pulse hurried. Rarely could he obtain an hour's continuous sleep at night. After the first mesmerizing he slept better than he had done for weeks, and in three or four days the continous fever was exchanged for a hectic attack, which came on periodically, about six o'clock in the evening, but rarely lasted more than two hours. The furred tongue became pure as an infant's, the appetite excellent, and so continued until within three days of the closing struggle.

" The pulse, also, sank into a calm state; which was only disturbed by the recurrence of the evening hectic. In all points, the improvement was so remarkable that every one about the patient thought that, had Mesmerism been earlier resorted to, the patient's life must have been preserved. But the abcess had already too deeply affected the bones of the back and hip, and had extended too near the femoral artery, to be checked by human means; or even by those almost superhuman aids which Mesmerism may be said to afford.

"Dr. Ennemoser attended his patient always at one certain hour (two o'clock) every day.

"During the manipulations, the sufferer invariably slept, but not always with the same degree of intensity. Even in the slightest stages of Mesmeric slumber, it was remarkable that R. T. could bear to have his right leg and thigh moved out of their customary position, and even somewhat roughly handled; whereas, at other times, an accidental touch upon those parts would cause him to shriek with agony. At all times, and in any moment, the Mesmeric passes either soothed or arrested the patient's sufferings, and having once been Mesmerized by Dr. Ennemoser, he was susceptible to the Mesmeric action of those with whom he lived in daily relationship. His brother, especially, who had come from England to attend his death-bed, could exert this influence over him, and had the ability to quiet his paroxysms of pain, or to make him sleep, at moments when under ordinary circumstances, any approach of slumber would have been impossible.

"What a valuable power was this! As any attempt at opiates increased the fever, and brought
on slight delirium, let any one imagine what un-
told treasures resided in the simple motions of the
hand, which calmed the thrill of agony, and gave
repose to the throbbing, yet exhausted nature!

"They who have heard, as I have heard, the
dreadful shrieks, the sounds, more resembling the
bellowing of a wild animal than the intonation of a
human voice, which are wrung from the poor sufferer (and that sufferer a much-loved friend), under
that most horrible, most appaling malady—the Lumbar Abscess—they, and they alone can appreciate the
almost overpowering thankfulness which swells the
heart in return for a gift that, in a few moments,
causes the shrill cry to sink into a tremulous murmur,
that murmur again to become an almost inarticulate
sob, and that sob to die, at length, away into the
blessed stillness of a deep restoring slumber!

"Oh we may reason and reason about the truth
of Mesmerism—the more or the less of faith we
bestow upon its wondrous, yet shifting phenomena
—but in such a case as this we FEEL, and in childlike spirit, acquiesce in the benefit, which we are
unable to analyze. "The heart is wise," as Southey
has beautifully said ; and here, at least, is demonstration in its glow of gratitude—that this power
is of God!

"These tranquilizing effects of mesmeric influence were manifested even unto the end of my
dear friend's life.

"Yet he owed a deeper debt than this to Mesmerism ! It had reclaimed him from the hardest
infidelity ! Of a singular organization, R. T.—the
most amiable of human beings—approached the
nearest to an atheist of any one I ever met with.
He seemed to want the very faculty which says at
once—a God must be ! But in his last illness then
it was that a new principle supplied the defect of
the original nature, more strikingly than if that
nature had from the beginning appeared full of
holy veneration. Who that then saw him leaning
over his bible, as he sate for an hour or two in the
evening, propped up by pillows on every side, calm
even under the attack of periodical fever ; triumph-
ing over mortal infirmity and pain ; rejoicing while
we inwardly mourned ; and whispering patience
and comfort to all around him ; who that beheld
this strength made perfect in weakness, but must
have said : " The hand of heaven is here !"

"And this faith—this wondrous patience—this
'holy comfort springing out of tears,' were (as he
himself told me), attributable, under Divine Providence, to Mesmerism.

"From having seen mesmeric phenomena, to
which he could not refuse his assent, he was led,
step by step, to recognize the mighty truth of
spirit predominant over matter—consequently of a
ruling spirit creating and sustaining all things. He
who sees and ministers to the spiritual wants of all
His reasoning creatures, had adapted this remedy

to the peculiar and fearful disease of the intellect under which R. T. had labored.

" And, touchingly, he said to me, 'I rejoice that Mesmerism should be the last remedy tried on me ; that it should prove successful in calming my pains, because it was the first thing that, through God's blessing, relieved me from the worse evil of an unbelieving heart.'

" And so, praying and praising God, and grateful for 'Mesmerism, the gift of God,' his spirit was loosed from its earthly bonds at the very moment when, after a severe paroxysm of pain, his brother was mesmerizing him, as he thought, into the calmest slumber. In connection with the relief which Mesmerism afforded to the last hours of my suffering friend, was another remarkable instance of the beneficial effects of the agency in question.

" The surgeon who attended R. T. in a purely surgical capacity, was an intelligent man, possessed not only of much skill in his immediate profession, but of considerable medical knowledge. Having frequently witnessed the power of the mesmeric passes to calm R. T., even under the severest paroxysms of pain, Herr Lorenz requested me to give him some instructions in the mesmeric art, which he purposed occasionally to employ, should a case, to which the agency might seem applicable, occur in his practice amongst the neighboring poor, many of whom he attended gratuitously. One day he came to me filled with pleasure and excitement, and related that he had had extraordinary success with Mesmerism, in a case which, for some days, had baffled his medical efforts, and had given him considerable uneasiness. The patient was a young man—a poor laborer's son, who was suffering under the worst form of typhus fever, and had been for a long period without sleep, talking incessantly, in a state of constant delirium.

"After about half an hour's mesmerization, the young man ceased talking, and soon after, fell into a calm and critical slumber. This circumstance so astonished the youth's grandfather, who was at the time in the patient's room (the common room indeed of all the family) that he retreated behind the stove to mutter Ave Marias and Paternosters, in order to exorcise the Evil One, who alone, as the old man thought (herein reasoning no worse than many a pious Protestant) could have possibly produced such a rapid change for the better. Fortunately his holy adjurations were insufficient to chase the kindly spirit that was calming and healing his poor grandson; and the young man awoke from the long and refreshing slumber which Mesmerism had induced, to an amended state, and to the possession of his senses.

"Under continued mesmeric treatment, the patient recovered his strength and health, as Herr Lorenz informed me, more rapidly than is usual with persons who have been similarly affected.

"A brother of the young man, and the grandfather, now fell ill of the typhus, the infection of which, being in the same chamber with the invalid, they could not escape. In these latter cases, Herr Lorenz assured me, the disorder was robbed of its horrors by a timely resort to Mesmerism, the employment of which was no longer suspiciously viewed by the old man. In fact, by a common reaction of feeling, he had for some time transferred the miracles of Mesmerism from the enemy of mankind's account to the credit of the Blessed Virgin.

"That the grandfather, who was eighty-three years of age, should have recovered speedily from this malignant fever, with scarcely any other remedy than Mesmerism (for in all the cases, but little medicine was given) is another proof of the curative powers of this influence. The instances of the utility of Mesmerism that have been detailed to me by foreign mesmerizers, some of whom are medical men—others not—are far too numerous to form part of a work; they would if duly set down constitute a work of themselves. I shall, therefore, only avail myself of the testimony touching the benefits of Mesmerism of one or two persons, on whose accuracy, from the coincidence of personal observation, I can place the firmest reliance.

"Dr. Ennemoser has assured me that he has successfully employed Mesmerism—either alone or conjointly with other remedies—in almost every ailment to which the human constitution is liable; but more especially where it was chiefly desirable to excite a healthy action in the nerves and brain. He considered this as the one appointed remedy for epilepsy, and declared that he had cured by it several cases of madness.

" Dr. Wilde, of Berne, informed me, that by mesmerizing a lady, suffering under periodical attacks of St. Vitus' dance, at the time when the affection used to come on, he succeeded in putting off the attack every day to a later hour, until its recurrence was altogether prevented. He also related to me a most interesting case of another lady, in the last extremity, under an attack of the Iliac Passion, whom he supported in his arms, every moment expecting her dissolution, when the happy thought came, like an inspiration, to his mind— try Mesmerism ! He did so, and the relief was immediate. The spasmodic action subsided, and the patient's life was preserved.

" During a residence at Dresden, I had the pleasure of hearing many authentic particulars of mesmeric cures from Count Szapary, a nobleman, of one of the oldest families in Hungary, who has devoted his fortune, time and talents to the exercise of Mesmerism, for the relief of disorders alone.* I have seldom witnessed a more touching festival than was a public breakfast given to this gentleman by the friends of those persons whom he had either benefited or cured by Mesmeric means.

" Many of the patients, now restored to health, were present at the table, and many of their relatives rose, one by one, to return thanks to the noble mesmerizer for the good which, under Divine Providence, he had done to those who were dear to them.

"One aged man spoke a few grateful words, but his voice faltered, and, tears choking his utterance, he was forced to sit down in a silence which was more eloquent than speech. It was for the preservation of his only daughter, snatched by Mesmerism from the very brink of the grave, that he desired to express his gratitude!

"A young lady, who sat next me at breakfast, a charming person, well known in the higher circles of Dresden, gave me a most interesting account of her own restoration to the blessings of health by the Mesmeric aid of Count Szapary. For two years, she told me, she had been unable to rise from her bed, or scarcely to move her limbs from nervous paralysis. 'At present,' she said, 'I can walk, ride, dance as much as I will; and many persons here, who know my former state, will scarcely believe in my identity with the poor bed-ridden creature that excited everybody's pity.

"'Some who meet me out walking, stare at me, and start as if they had seen a ghost. Not, indeed, that I look much like one at the present time;'— a remark in which I fully coincided.

"Indeed, the peculiarly blooming appearance of this young lady so much belied all past suffering, that, till I entered into conversation with her, I never dreamed that she could have been a mesmeric patient. The efficient cause of her cure, this lady declared to me, was a mesmeric sleep, which lasted a fortnight, of the events of which, she positively assured me, she had no manner of recollection.

"Yet, from the testimony of her parents, and of those who were about her, during the period of the slumber, she could not doubt that she had in this time spoken, taken nourishment, and beheld objects, although her eyes were closed. Such lengthened mesmeric trances as this were frequently, as I heard from Count Szapary himself, and from others of his patients, produced by him in certain cases, and were a powerful means of restoration to health. Not, as the Count assured me, that he labored at producing any such effect. On the contrary, he believed (and Dr. Chapelaine, in Paris, told me the same thing), that the mesmeric sleep was by no means an indispensable condition of mesmeric benefit.

"Many of his patients, he said, having got well under daily mesmerizing, without having experienced any drowsiness, or extraordinary symptom whatever. His great desire was to leave Nature free, to select such means of renovation as should be most suited to her exigencies—to follow where she guided, not himself to lead the way, and to concentrate his thoughts as much as possible, upon the single, simple idea of doing good to the patient.

"But where the trance occurred, as it did mostly in exhausted, yet irritable temperaments, the Count considered that this was the one thing wanting to restore the equilibrium of the constitution, and to enable Nature to recover her tone by a prolonged repose.

" That Count Szapary should, as it were, involuntarily cause this effect, so rare to be used by other mesmerizers, is a proof of that which I have always believed, namely, that to each person belongs a mesmeric action, as peculiar and as individual as his own character. Count Szapary certainly possesses great mesmeric energy, being able to mesmerize many persons in the course of the day, without fatigue to himself.

"In England, not less than on the continent, opportunities have been afforded me of gathering evidence respecting the good that Mesmerism has effected.

" I have accompanied Dr. Elliottson in his visits to such patients as he was treating by Mesmerism, and have listened to many an artless expression of gratitude, poured out to him by persons of various classes, for the important benefit they had derived from this calumniated remedy.

" One of the most eloquent of these was a respectable woman, with whom I found myself accidentally in the same room at Dr. Elliottson's while waiting one day until he had leisure to see me. This person assured me, that for years she had suffered a martyrdom from tic-douloureux, to such an extent, that when Dr. Elliottson first began to mesmerize her, she could not hold any cold substance, even an open book (unless it were previously warmed at a fire) near her face, without a paroxysm of pain, which would send her darting involuntarily from one end of the room to the other.

Of course, the contact of the open air would have been unbearable torture to her. "And now," she said, "I walk every day, in all kinds of weather, to Dr. Elliottson's house, without inconvenience; and instead of having to suffer many attacks of agony in an hour, I often pass a whole day without pain.'

" There was something in my accidental meeting with this woman—in her earnest and natural story —in the utter absence of pretension to display, which impressed me strongly.

"And this is only one case of many such. I am indebted to Mr. Wood, who has studied under Dr. Elliottson, and who also admits Mesmerism into his medical practice, for having witnessed one of the most triumphant proofs of the power of this influence that can be conceived.

" It was the case of a little boy, only three years of age, who was born an idiot, and who had distressed all those about him by incessant shrieking, especially during the night; in consequence of which his mother declared she could scarcely obtain any sleep or repose. The mother, a respectable person of the middling class, brought in the child, which appeared not larger than an infant of fifteen months old; but no infant of that age, to whom the gift of reason was vouchsafed, ever presented that fearful absence of expression, which, in the dropping of the lip, and unmeaning eye, characterized this unhappy sufferer. At the time when Mr. Wood began the mesmerizing, the child was keeping up a sort of low moaning and occasionally crying aloud in a strange and disagreeable tone of voice.

It was also moving restlessly about in the mother's arms.

"After about five minutes' mesmerization the change in the child's demeanor was most striking. There was no longer any moaning or restlessness; and the eyes had gained a human look, that grew upon them till they began to close in slumber, which at the end of little more than ten minutes, was complete, and of the deepest kind. At this time the countenance had lost everything that was unpleasant.

"It had even the remarking and attentive expression which so strongly distinguishes the mesmeric from common sleep, so that, in gazing upon it, one could now say, 'There is mind here!' About the eyelids, especially, was that singular look, so peculiar to mesmeric sleep-waking, as if the eye were looking through them to something beyond.

"I was allowed to test the depth of the child's slumber by rather rough methods, such as shaking it, shouting out loudly close to its ears, &c., but nothing disturbed its repose in the slightest degree. The arm was taken up, and suddenly, left to itself, fell like lead, and more lifelessly than in ordinary sleep, upon whatever arrested its descent.

"The mother now placed the child upon the bed, and assured me that it would sleep there quietly for hours, and awake in a far more tranquil state than that in which I had first seen it. The improve-

ment in the child's condition, since the time when
the mesmerization was undertaken, she told me,
consisted in the following particulars: The rest at
night was now good, and, for the blessing, she said
she could not be too grateful.

" They alone could appreciate it, who had known,
like her, what it was to sit up, night after night,
with a child in that fearful state of idiotism.

" By day, the paroxysms of crying, and of convulsive motion had become less frequent, and of a
milder character; and, whereas formerly the child
took no notice of anything whatever, it now began
occasionally (and more especially when awakening
from mesmeric slumber), to show symptoms of being interested in various objects.   But, above all,
there was an evident dawn of speech, and the
words 'papa' and 'mama' were, however imper-
fectly, uttered ; but, before Mesmerism was tried,
nothing approaching to articulate sound had ever
passed the child's lips.

" I do not instance this as an evidence of the
power of Mesmerism to restore confirmed idiocy
to reason—in other words, to work a miracle.

" The child was relieved, not cured, and is, I believe, an idiot still—though of a milder kind than
formerly.

" Afflicted as he was with a radical disease of the
brain, a perfect restoration of the mental functions
was not to be looked for ; but greater progress
toward reason had perhaps been made could Mr.
Wood have concentrated his time and mesmeric

power upon this one patient ; but he was called out of town to attend upon an urgent case, which occupied him for some weeks, just at the time when he was daily producing more remarkable effects upon the little idiot boy.

" This statement will probably nullify all that I have said regarding this extraordinary case, because, with most persons, not to cure is equivalent to doing nothing at all. But I must say a few words on this point.

" Is mitigation of suffering nothing in this suffering world ? Who are the persons who choose to deny the advantage of a remedy because it is not entirely curative ? Certainly not those who are tortured by disease, nor their friends who witnessed that torture. It is the partisans of a system—the prejudiced—whose prejudices no attack of serious pain, neither of illness, has shaken ; who, being in possession of health themselves, can very calmly imagine that a pang more or less is of but very trifling consequence ; and that, more especially, any one pang rendered less by Mesmerism, must be treason to all that is regular and respectable in this world of words and forms. To me the case of the idiot child, however stopping short of perfect relief, is deeply interesting in sundry points of view. To produce any mitigation of symptoms, any glimmering of reason—in a case of born idiotism—is more extraordinary than to effect a cure of delirium, or mania, springing from accidental causes. This partial triumph over organic defect shows Mesmerism capable of producing, to the verge of the possible, and of educing from certain elements all the favorable results of which those elements are susceptible. The fact of Mesmerism having been able to compose, and so quickly bring into slumber, a child of three years old—that child, moreover, an idiot, suffering under high irritation—is sufficient of itself to refute the opinion of those who would refer Mesmeric efforts to the delusions of the imagination."

The thoroughly practical character of Mr. Townshend's facts and deductions, justifies the extended quotation I have made from his interesting work.

The statements are the more valuable as coming from a thoroughly disinterested observer, and noticed by him as mere incidents, rather than as primary facts related to the field which he had prescribed for himself in his investigations in Psychology.

They also have value by giving a *practical* answer to some of the weak objections of weak minds, who in this day, as in the day for which he wrote, are carried off their feet by the extraordinary character of some of the cures effected by this agency, and tremble lest they have stumbled upon something which shall undermine their Christian faith, or touch upon some forbidden ground which has heretofore been looked upon as miraculous and supernatural.

In the earlier history of the science some weak-headed clergymen felt called upon to denounce the practice as an emanation of evil, just as they preached sermons against vaccination upon its first introduction. All these objections have died out, and it would be a waste of time and patience to travel over the ground now, as these cavils have long since been fully answered by deeds of *good*, instead of the *evil* effects which were predicted by such opponents. If, however, the reader would like to follow a polemical discussion he will find an exhaustive one in the Rev. Mr. Sandby's work, entitled " Mesmerism and its Opponents," or in a work by Mr. George Barth of London, styled " Mesmerism Not Miracle." These works are devoted to this class of objections, and fully answer every imaginary cavil of this character.

One such testimony as Mr. Townshend adduces, where his friend's infidelity was made to give way, and his mental darkness to flee with his racking pain, while he exclaims: " Mesmerism, through God's blessing, relieved me from the worse evil of an unbelieving heart," is a better answer than any amount of dry speculation, or angry denunciation, and vindicates that wise saying that "a house divided against itself cannot stand."

The curative power of this agency stands where it belongs, *within the circle of natural law*, and is in no way in competition with the grand displays of Omnipotent power made by our Divine Master when He was on the earth.

XIV.

I have now reached the limit prescribed to myself for the presentation of facts and statements relating to magnetic cure.

The great difficulty has been to select from the huge mass those items that would command most general attention from the American public, and convey an adequate idea of the magnitude of the investigations prosecuted in Europe. It is to be hoped that the wave of inquiry shall yet sweep with irresistible force across this land, bringing into activity powers for good that now lie dormant and unused.

Only those medical men who have had access to foreign authors, so difficult to obtain in this country, have any *technical* knowledge; and the number who have had personal experience in the practice is still less. This is evidenced by the crude attempts to discuss the subject in medical conventions, where the vast majority must sit dumb, and frequently those who present its claims seem ignorant of the fact that the same things have been matters of common intelligence abroad for more than a quarter of a century. I trust such sources of information will be opened by the perusal of these pages as will lead earnest minds to seek closer contact with the thought and experience of the authorities to whom I point, who have already done so much in the field that many imagine to be so new.

I claim no originality of discovery. I have proved by experiment and practice much that is set forth by the authorities presented to the reader on this branch of the subject ; and while I have not been indifferent to the fascinations of the so-called higher mental phenomena often met with in magnetic practice, I have studiously avoided introducing this element into what I design to be strictly a review of the subject in its relations to the healing art, believing that anything beyond this, at present, will be but confusing and out of place, and of no practical value.

For the same reason I have omitted the discussion of processes, upon which so much has been said by others. No mere *form* of inducing the magnetic condition has pre-eminent advantages over all others, and there is no solid basis, in my judgment, for any such distinction; what is claimed for one condition is often misapplied in others, and each process doubtless has its advantages under given circumstances. What may appear simple enough in itself is not a proper thing for indiscriminate use by the thoughtless and ignorant, who can alone do mischief with a power that should be sacredly guarded. There are plenty of avenues open for the acquisition of knowledge in this direction to those who are entitled to be in-

/ formed in regard to it, and I do not propose to be one of the many who have encouraged indiscriminate, useless, and foolish, if not dangerous, experiment. To the elucidation of this folly a few paragraphs upon SYMPATHETIC INFLUENCE may be devoted by way of caution to the ill-informed.

XV.

SYMPATHETIC INFLUENCE.

This is a subject that concerns practitioners more than patients, and while there are widely varying susceptibilities and idiosyncracies among patients, the same is true of the practitioners, and though many of them may escape sympathetic effects from contact with numerous diseased conditions, others experience more or less discomfort, and are perhaps exposed to some personal risk.

The writer himself has often experienced temporary effects lasting for an hour or two, but never leaving any permanent traces behind.

It is doubtless true, also, that active and successful practitioners are most liable to a disturbance of this kind, but if they have a strong will, and powers of concentration, it can be readily overcome.

The absence of these qualities probably accounts for so many men failing, after a brief practice, in this profession.

These effects and the natural ability to overcome them, must be settled by the experience of each practitioner for himself ; no rule can be laid down of general application. Many practitioners have lived to a good old age whose labors were tremendous ; as, witness Doctor Mesmer, who lived to be over 80 years of age, his own last illness being delightfully soothed and ameliorated by vital magnetism.

In this line Dr. Reynoldson, of Bootle, near Liverpool, reports, November 8th, 1847, on page 325 of 5 Vol. *Zoist :* "Having for several years devoted myself entirely to the cure of disease by mesmeric action alone, I have been daily, and often hourly, conscious of sympathetic pain when mesmerising a patient ; and I notice this pain occur the same minute that the correspondent pain leaves the sufferer, the pain generally passing away in a few moments or minutes, but occasionally continuing with much severity for some hours, and once it continued about the diaphragm for eight days, and was then relieved by magnetism." Again he says : "In my practice here, upon two occasions, I have been banished from my patient in fever by the advice of physicians."

" The one patient turned dark in typhus fever and recovered; the other soon recovered. But my nights were sleepless for hours during the progress of each case, and I can testify from most painful experience that intense suffering may be inflicted upon the mesmerist, who, after establishing a strong sympathy with his patient, is not allowed to continue to manipulate during the severe stages of the disease."

Dr. Elliottson observes :

" The fact of the transit or metastasis of disease from one individual to another, whether of the same or of a different species, is unknown to the medical world. But this pathological occurrence is proved by Mesmerism as clearly as the physiological fact of cerebral sympathy of sensation and thought between two individuals. I am not aware that any account of it exists in medical works, but probably very many instances are recorded in mesmeric writings, since it is commonly spoken of amongst mesmerists."

The following case in point is also furnished by Dr. Elliottson, supplemented by a letter addressed by the patient himself to the editor of the *Zoist*, and found upon page 236 of the fifth volume of that journal:

" A young tradesman in my neighborhood, whom I had never seen, came to consult me on the 25th of August, 18.46; he had dark hair, and was pale and slim, but possessed of remarkably firm muscles, and was so swift that he had beaten the most noted runners. He informed me that he had been ill more or less for some years; that the slightest circumstance had cast him down or irritated him, according to its nature; that he felt sometimes as if he should go out of his mind, and sometimes as if he were about to die; that he slept well, but was tired before going to sleep, and awoke in the morning tired, though he had no muscular weakness; he frequently experienced a heaviness of the eyes; sometimes numbness and tingling all over him; sometimes violent itching of his arms; his hands were cold and shriveled; the bowels were often torpid, and at those times he was always worse; that the liver secreted too sparingly, and that he was better in the country. His mother was nervous.

" I prescribed in the best way I could for him. In about ten days he called upon me one evening, with his symptoms very intense, and suffering a headache so severe as almost to drive him mad. This lasted till next morning; and when he called again in two days, he was in the first state as when I first saw him. On the evening of this day I left London for Switzerland, and did not see him again till the beginning of November.

" After being under my treatment for two months, he was no better. I had not mentioned Mesmerism to him, although I believed from the first it would be the best thing for his case, because I have been compelled for a long while never to name this subject before inquiries are made to me. I found a large number of medical men, physicians, and general practitioners equally, who were perfectly ignorant of it, who had done their patients no good, and had no hope of doing them good, stare at me with a far more sagacious look than I could give, and assure me they considered it nonsense, and would not consent to its use; and I found large numbers of patients refuse to hear of it, because they knew it was nonsense, dangerous, or Satanic. I give my opinion honestly when it is asked, but I will never run the risk of finding persons unreasonable or ridiculous, and of being thwarted in an attempt to do my duty by them.

"This patient and his wife had heard of Mesmerism, and now, as he was no better, inquired what I thought of it in his case. I immediately informed them that it ought to be tried, and was very likely to be of essential service."

Dr. Elliottson instructed the wife of this patient so that she could continue the magnetic treatment in his absence. It resulted in the man's complete recovery, but also in the transference of the disease to the wife, who was the operator. Their mutual experience was communicated to the *Zoist*, as follows:

To the EDITOR of the *Zoist :*

77 Wells street, Oxford street, *Sept.* 3, 1847.

SIR: I feel it to be a duty I owe to suffering humanity to make known my experiences of the powerful agent to whose salutary influence I am indebted for the healthy mind and body at the present moment engaged in writing this narrative of my relief from mental and bodily torture. Yes! against the senseless raillery of its opponents, I will record my grateful testimony to its virtues. And at this moment—with a mind undisturbed by phantasies, and a body in which the stream of life is flowing redolent with health—I might be pardoned, were the pleasures of my present existence contrasted with the bitterness of the past, to hurry me into the language of enthusiasm. Such, however, is far from my intention ; I wish to give a plain statement of my sufferings and cure, in the hope that some poor sufferer may be led to disregard the nonsensical opposition of bigoted ignorance, and use the means which by the blessing of God proved so healing to me.

I was twenty-seven years of age, the son of affectionate parents, the husband of a devoted wife, the father of healthy children, and in a position of life insuring most of life's comforts.

And yet, with all these means of happiness, I was a wretch, to whom the years of life were years of misery—imaginary if you will—yet misery ; my smile was an effort to prove myself not divested of the sympathies of humanity ; the dread of suicide followed me as my shadow, while from my temples scarcely ever fled the *dull, dull* pain, which sometimes increased to positive agony. To my morbid fancy, no security could preserve my children from untimely deaths, no precaution ward off evils whose dark shadows blackened my soul. These symptoms softened at times, yet ever preying on my peace of mind, were my companions for years.

My temples are indented with leech bites, my arms scarred with issues, my body was poisoned with medicine. I never recollect feeling myself what I could emphatically term in health, until the last few months, in which I have taken greater liberties with myself than I ever did in my life before.

It was after one of my most violent paroxysms, I had the happiness to consult Dr. Elliottson, the indomitable champion of mesmeric truth. The man who, for a great principle in nature, of which he was convinced, defied the sneers of ignorance, the malice of envy, and the filth of misrepresentation ; and who, after suffering, as he must have done, a martyrdom of anxiety, has lived, thank God, to see it fully borne out by the experience of thousands, and—the only reward he sought— seen its solitary influence exhibited in the alleviation and eradication of human suffering.

Excuse this digression ; it is the outpouring of a grateful heart, warmed by the recollection of its benefactor. By the advice of Dr. Elliottson I was induced to try mesmerism, and my wife having been shown the way to exert its influence upon me, commenced operations.

At length, about a month after the first essay, I sank into a mesmeric sleep.

Its influence from that time was complete ; a few moments only served to send me into a slumber, from which I awoke a healthier man. My cure was rapid and complete ; physic was thrown to the dogs ; and any recurrence of my former symptoms vanished at the wave of her hand. Then it was that a most singular phenomenon was shown.

I looked upon my wife and beheld in her an image of my former self. The ills which before I had to mourn over in myself, I saw in her. The same dejection, the same feeling of weariness, the same pains. The high-spirited, light-hearted woman turned into the desponding invalid. I had bequeathed to the mother of my children the malady which had made my life a misery. But who under such circumstances could give way to despair ?

Mesmerism had expunged the word from my vocabulary.

We waited upon Dr. Elliottson, who kindly mesmerized my wife daily ; and in less than a fortnight I had the happiness of seeing her the same happy creature as heretofore.

I have had but one return of my old enemy, was mesmerized, and awoke well.

Such, sir, are the particulars of my case. To any one seeking relief from pain, I shall be happy at any time to render any information in my power ; and the happiness I experience

.n my recovery will be enhanced, if I can be assured of one
)oor sufferer following my example, with, I feel confident, the
ame happy results.

I am sir, your obedient servant,

WILLIAM SNEWING.

These instances will suffice to give fair warning
to inexperienced persons of the risk and discomfort
they may incur by tampering with a force they do
not know how to control.

XVI.

DR. NEWNHAM'S EXPOSITION OF PRINCIPLES AND
PLEA FOR INVESTIGATION AND ADOPTION BY
THE PROFESSION.

The practice itself need not and ought not to
come into *competition* with medical practice of any
school, nor to exclude medical science in any form,
but should be accepted as one of the physician's
most powerful aids, for its advocates and expounders have been found in every school of medicine,
and among their most eminent representatives.

While it will be seen that its range is exceedingly
wide, its proper field of application is doubtless

more specifically within the line of nervous diseases where medication has failed to afford relief.

I plead for unbiased investigation, and to this end let that eminent English surgeon and writer, Newnham, speak to his brethren :

"It is important," he says, "that magnetism should be investigated as other sciences ; that acknowledging the existence of the agent, the modes of its action should be ascertained, and that the inquiry should be conducted by men whose scientific habits will enable them to collect, to weigh, to discuss, to estimate, to discriminate testimony, and to adopt or reject it, according to its intrinsic merits. It is, however, to be remembered that even here is a concealed evil which must be guarded against.

"While scientific men are generally considered to be free from *vulgar* prejudice, it is a fact that they are liable, in no small degree, to a class of prejudices peculiarly their own, and marked, where they do exist, by an intensity and pertinacity which are wanting in feebler and inferior minds. Prejudice in the vulgar and uneducated, or partially educated, commonly assumes the form of all-devouring and immeasurable belief, and a yielding up of themselves to the influence of feeling. With the learned, on the contrary, the very antipodes to the former class, prejudice assumes the shape of skepticism, a doubt as to everything which is not proven to demonstration, or which appeals in any way directly or indirectly to the voice of feeling or passion, or to the influence of imagination; it is reason alone which is to be attended to, and an amount of proof is asked for which is incompatible with the imperfect state of our knowledge and faculties, of our acquired stores of wisdom, and of our powers for further acquisitions.

"It is one of the prejudices of great minds (for great minds are quite compatible with many little-nesses) to dislike novelty because it is *new*, and that, perhaps, precisely in the same proportion as this characteristic is attractive to little minds ; and therefore, since the phenomena of *produced and lucid* somnambulism are rare, they have been invested with more than usual doubt. Hence the very great importance, in order to arrive at *clear* results, to distinguish between the two from the beginning, and not to confound MAGNETISM, whose existence is shown by incontestable proofs, with *Somnambu-lism*, whose features are of a more questionable and protean character, and is a state oftentimes *indepen-dent* of, but occasionally developed by, magnetism.

" The truth is, that in tracing the history of the relation of facts to their causes, and between them-selves, observers have often been mistaken, in their degree and mode of *relation*, either from their being mere *theorizers*, and from their determination to theorize, before they inquired, or that they had not the required amount of knowledge, to enable them to judge accurately or that they were carried away by their feelings. In order to avoid *all* these errors, it is necessary first to aim after impartiality,—to scrutinize the narrative of facts, to test what is doubtful, to separate the true from the false, and to endeavor to arrange the residual product under some known physiological laws, always taking care not to confound the cause and the effect *as one*, and thus to argue in a very limited, but never-ending circle of truism. The *facts* should be compared with other facts, judged of, and reasoned upon; but the reasonings must not precede the careful *record and observation* of facts. It may be asked, perhaps, *what is a fact?* This question will be answered by the ignorant, as everything which their credulity can believe ; and by the learned, it will be defined to be anything within the limit of their *previous knowledge.*

"Both answers are fatal to improvement, yet there is more hope that the *ignorance* of the vulgar may be cleared away, than that the *intolerance* of the learned should be *overcome;* the former may perchance be accessible in instruction, but the learned invest themselves with the supercilious panoply of their own pride, and despise and condemn everything beyond the reach of their ordinary views.

"It is, however, the part of puling credulity alone to believe everything which may *seem* to be, even under such favorable circumstances; and it is the part of half-informed scepticism to deny *apparent facts,* when we are so ignorant of the resources of nature, and when she has so many ways, to us perfectly unknown, in which may be concealed the truth, for which we are professing to inquire.

"To escape from this ignorance on the one hand, and presumption on the other, it is right to withhold our opinion till we have dispassionately inquired into the facts and circumstances, as well as into the reasonings upon them, of those who have preceded us; and until we have done so,—though we may prudently array ourselves in scientific *doubt*,—we shall, if guided by the same principle, take care to avoid that amount of *incredulity*, which though it be productive of less injury to science than its antagonist principle of a too easy belief in the marvelous, is still so serious an error as to demand our utmost care to escape from its entangling meshes. The result of the judicious balancing of these principles will be the development and establishment of general laws, which can only be drawn from the observatio and accumulation of facts, sufficiently numerous, and adequately observed.

"But it will be said by an objector, that to believe in an extraordinary fact at once is only the part of weakness and enthusiasm ; that in proportion as a fact is *extraordinary*, so will be the probability of error or falsehood on the part of the witness; that a greater amount of testimony is required to substantiate a suspension of Nature's laws, or any extraordinary development of her previously unknown resources: moreover, the very things which *add to the strength* of belief in unenlightened minds, are precisely those which *detract from it* in the minds of the well informed. The one being greedy after the marvelous; the other always suspicious of its reality; and that, therefore, it is impossible to sacrifice our common sense, and the general report of ages, at the gratuitous bidding of a few magnetizers.

"We rejoice to reply to an objection thus candidly stated. We admit that the exhibition of too easy credence is the mark of an *enthusiastic and a feeble mind,*—but we fearlessly assert that obstinate belief without examination is the evidence of an *unsound mind.* We have before shown that such is the character of the witnesses as to the facts, and such their agreeing testimony, as to render error to the last degree *improbable*, and falsehood *impossible.* We have shown that an over-abundant array of testimony in favor of the facts of magnetism can be found,—even if these facts were miraculous, and involved a suspension of the laws of nature; but we have still further shown, and shall hereafter more distinctly prove, that they do not involve such deviation or suspension ; but that on the contrary the facts are deeply laid in the hidden things of nature; that they are recorded as natural and spontaneous products and newly exist in the shape of magnetic facts only as the development *at will* of their hidden resources, with the laws of which we are still ignorant. Moreover, we contend that the facts are not miraculous, though at present they may be inexplicable, and finally, we do not ask of any one the sacrifice of common sense, but only the *adaptation of that common sense,* without prejudice or prevention to the sober investigation of the facts. In the practical inquiry into these facts, it is not sufficient to employ the magnetic processes ; for as the results may be *nothing*, or *useful*, or *hurtful*, according to the individuals operated upon, or according to those conducting these processes, and according to other circumstances and things, which cannot always be foreseen, because our acquaintance with the physiology of the nervous system is yet in its infancy, it is quite clear that the inquiry should be directed exclusively to legitimate objects on the one hand, and to those experimenters on the other who have the judgment to discriminate, and the probity which will enable them to be firm in principle.

"Such sincere investigators will always be rewarded for patient attention ; but the CURIOUS spectator will never be convinced by the exhibition of phenomena, which contradict his preconceived opinions ; and even the *philosopher* "convinced agaist his will," remains "*of the same opinion still.*" There is no good without its corresponding evil, and the characteristic of the age in which we have rejoiced, and do rejoice, viz.: that of seeking after *facts*, has also tended to limit our researches after knowledge, and an enlarged acquaintance with the boundaries of scientific truth. For these *facts*, as they are called, are really the effects of some one or more *concurring cause ;* and the minds of men which have discovered phenomena, the cause and nature of which extended beyond the boundary

of their *former notions,* have been too frequently satisfied with throwing them aside as unworthy of
further investigation, or as having already reached
the utmost bounds of knowledge. And yet, however much this *might be* allowed of phenomena
with which we presume to be sufficiently acquainted,
it is to the last degree *unphilosophical* and intolerant with regard to those of which we are confess-
edly *more or less ignorant.*

"As yet, with respect to magnetism, there is much
to learn; and it presents so many *apparent* anoma-
lies, whose causation seems to be entirely beyond
the reach of any of the known laws of nature, that
they who have studied it most assiduously are still
not safe from the apprehension of mistake.

" Nor is this wonderful; nor does it derogate one
atom from the character of the pursuit, or of those
who investigate.

" Medical men, of all others, should be the last to
avail themselves of this undefined and uncertain
state of things; for let them fall back upon their
own peculiar science, and let them say if they can
explain *all the facts* which present themselves to
their notice.

" In truth, we must be thoroughly acquainted with
all the mysteries of creation, before we can dare
to pronounce on the possibility of one fact, and the
impossibility of another; we cannot even generally
trace the causation between an organ and its function, although we may presume to be tolerably
well acquainted with both; and even if we were

able to do so, with regard to one organ and function, we should have no means of judging how far under other circumstances these may have some direct or relative influence upon some other organ or function in its sound or morbid condition. How then shall *they* unblushingly ask for a greater amount of certainty in the sequences, and of perfection in our acquaintance with the comparatively unexplored phenomena of magnetism?

"No wonder, therefore, that the converts to magnetism are among those who have *witnessed* its power, while its most virulent opponents have been those who have *never seen its effects*, who have obstinately *refused to investigate them.*

"One thing is certain, that when different persons, at different times, and in distant places, have arrived at the same results—and these results are supported by natural analogies, that there must be *some truth* in these novel doctrines—*some things positive* and *intangible*, notwithstanding the mass of error with which they may have been associated, may be so considerable as to render it difficult to distinguish truth from falsehood, especially if the investigation be abandoned by the wise and prudent and well informed, and be left to the ignorant, the prejudiced and the unthinking.

" Thus in the days which are passed, and especially during the middle ages, *miracles* (so called) were of frequent occurrence, which might easily be accounted for on natural principles. *Now*, we no longer speak of miracles; but if anything *extraordinary* occurs, we seek to examine first the truth of the facts stated, and then proceed to *explain, if we can,* the peculiarity of the attendant circumstances.

" Notwithstanding the progress of medicine, and its dependent sciences of late years, none can deny how utterly powerless are our remedies against a multitude of disorders, especially against those which affect nervous structure and function ; and for this best possible reason, that our acquaintance with both is so limited, so undefined, and so uncertain, that we are unable to form anything like rational indications, for the relief of their disordered conditions. The knowledge of a disease is more than half its cure ; but if we have almost *no* knowledge, it is certain that our power of administering relief will be very limited.

" Yet the highest object of medicine is to *cure;* and if, under such circumstances of acknowledged failure, any new remedy can be proposed, we should not disavow its assumed curative agency, and its consequent assistance in prosecuting this object, till we have proved its faithlessness, merely because we are unacquainted with its *modus operandi* upon that very structure and function with whose primal condition we are ignorant. Nor is there any reason why we should *refuse to investigate*, because among the facts alleged by magnetizers there are *those* which exceed the possibility of belief, among many others which are undoubted and undeniable ; that error *may* be *mixed* with truth, is a grand motive for inquiry, but no palliation of unbelief.

"Magnetism rests on a great number of real facts; but its adversaries have generally avoided the *real* question, and have preferred wasting their strength upon speculations as to the *nature*, rather than as to the *phenomena and effects* of magnetic processes.

"When, however, men of learning, of character, of honor, of high probity and professional standing, men who have no interest to deceive, and everything to lose by dishonesty—when men of such character attest the same facts from one end of Europe to the other, when they describe corresponding circumstances, and detail similar phenomena, how is it possible to throw the blight of willful doubt upon their assertions? How is it that medical men generally do so tacitly asperse their own fraternity, as they necessarily must do by refusing to inquire?

"True it is that a great surgeon once said, in disparagement of medical authority, 'Medical Facts! Medical Lies!!' But John Hunter was not a man well calculated to form a just estimate of professional testimony; the doctrines he propounded were not at first sight admitted, and his unhappy, irascible temperament was easily ruffled and irritated by opposition; and it was in one of these moments that he uttered the bitter and unjust sarcasm above detailed. But were it *as true* as it is unquestionably an exaggeration, it would not give

to the medical men of the present day a *locus peni-
tentiæ* for their treatment of the question of magnetism, since this also rests its truth on the testimony of the simple-hearted, unlearned, unprejudiced, and unsophisticated observer, as well as on
the finer-drawn observations and reasonings of
professional and other learned persons.

"It is, then, the paramount duty of the medical
inquirer to allow nothing to escape *his* notice which
may add to the perfection of his art, extend his
means of relief, or contribute to the solace of suffering humanity.

"Upon this general principle we take our stand:
we assert that, by many credible witnesses, magnetism is announced as such a means, capable of producing (under some circumstances) the most beneficial results; and granting for one moment, for the
sake of argument, that all these witnesses may have
been *deceived*, and that the whole miscalled science
is a delusion, yet even this supposition does not
warrant our indifference or rejection, because it
may be only that the facts have not been observed
with sufficient accuracy; and the *fact assumed*, that
many witnesses have been deceived, affords the
strongest evidence that the circumstances observed
have, at least, so much the semblance of truth as
accurately to represent its features to many inquirers; and if so, *what is it which gives that resem-
blance?* Is it not most probable that the system
thus observed does possess *some*, though possibly
exaggerated, truth? In such a position, to search

after and to ascertain this modicum of truth, is the part of wisdom and honesty of purpose.

" The forgetfulness of this great principle has led some medical men to avow their disbelief of magnetism, although they had not studied its phenomena ; and their being highly esteemed as men of science has not preserved them from falling into vulgar error, and, curiously enough, has reduced them to *present in their own persons* some of the higher phenomena of magnetism *while in the very act of denouncing them*, for they have fallen into a state of *waking slumber*, and *have become blind and deaf, and inaccessible to the general subjects involved in the inquiry.*

" This apathy, this cataleptic state of the intellect, has been most extraordinary, especially when contrasted with the ordinary proceedings of medical men. Under common circumstances, they do not hesitate to employ the most heroic experimental treatment, and to administer poison in any shape, but they dare not *conscientiously* place their hands upon their patient, with the intention of curing his maladies, and with a fervent and sincere desire to do so; and yet according to the father of medicine, *Hippocrates*, 'Nothing should be omitted in an art which interests the whole world, which *may be* beneficial to suffering humanity, and which does not risk human life or comfort.' "

So fair, so strong, so urgent an appeal as is this, for investigation into the facts of a science of which Dr. Newnham became one of the best investigators,

should come with special force to the medical profession, of which he himself was an honored member. When he commenced his investigations he
was strongly prejudiced against its alleged phenomena, and a disbeliever in its curative power;
and for this reason he was urged to write against
it, material being furnished him for the purpose by
others ; but, being an honest and fearless man, he
commenced *de novo*, and the result was a treatise
in its favor, instead of against it.

His work entitled "Human Magnetism" is a
splendid instance of the triumph of truth over
error.

Whatever may be the effect of such pleas on the
average professional mind in stimulating to an examination of the merits of magnetism as a curative
agent, it cannot be denied that it stands supported
by an imposing array of facts that are indisputable, and numbers among its advocates men who
stand as high as any in the ranks of science. ·

Fortunately those to whom the practical question
must come, are, in the main, free from the ob-
stacles which stand in the way of many medical
men, whose professional pride and self-interest may
cause them to shake their heads wisely, and doubt-
ingly deprecate the trial of anything with which
their experience has not made them familiar, but
which the fair, unanimous, and convincing testimony of thousands has shown to be safe and
harmless, when properly administered, and against
which *no evil consequences* have yet been proved, al

though predicted so often and so persistently during forty years of agitation and investigation.

To the invalid stricken with nervous disorders
involving brain and spine, causing long and weary
days of hopeless suffering, these pages will come, I
trust, like a new revelation, to inspire hope in the
hopeless and bring courage to the despairing.

To you I would say, magnetism will not perform
impossibilities, but it will cure in many cases that
are pronounced " incurable," and will give *relief* in
nearly all. To those who are sufferers from paralysis, epilepsy, sciatica, spinal disease, rheumatism, tic-douloureux, lumbago, neuralgia in its
various excruciating forms, sleeplessness, hysteria,
malarial troubles, and functional disorders that
may have defied all other means of cure—to you it
is mainly important to say that thousands have
found relief from such distress in nature's own remedy.

This fact alone is important, and you can safely
leave the professions to wrangle over their *theoretical* vagaries, whether advocates or opponents.

The evidence in favor of the practice is but
glanced at here. In the thirteen volumes of the
Zoist, so often referred to, over seven hundred cases
of a desperate character are given in detail, in all
of which either an absolute cure or permanent relief was obtained; if we add to this the vast mass of
private practice in England alone, which never
came to the public eye, we cannot but feel it to be
a strange thing, that at this late day there should

appear to be a necessity for pressing the claims of *vital magnetism* upon public and professional attention.

If by this *resume* of views and collocation of facts, experience and observations, I shall contribute to a more intelligent comprehension of the *utility* of this wonderful agent, I shall be content, trusting that abler pens will continue to diffuse right and worthy views of the question, and thus neutralize the mischief which the weak and foolish pretensions of some modern magnetists have wrought in assuming for themselves *supernatural* personal endowment, disgusting intelligent investigators and thus preventing them from pursuing the matter in a rational mannner.

XVII.

THE LESSONS OF HISTORY—SHALL WE BE STUPIDLY TIMID OR BRAVELY WISE?—A WARNING TO OBSTRUCTIONISTS.

In the future of magnetism there will doubtless be drawbacks, as there have been in the past.

The first physiologists of the day are speaking in terms of commendation of it as a means of cure;

but the rank and file of medical practitioners must necessarily feel bewildered for a time.

But, I would say to such as are in doubt, whatever may be your lack of information on this subject, do not range yourselves with those who decry it unheard, or because of its novelty.

There are but few advances in science or medicine that are not opposed and scouted by men of small minds, who are fearful that if the groove or rut in which they have been running is disturbed there will never again be found a place for their feet.

Mr. Lang, of Glasgow, in 1843, set some of these facts in a strong light in his reports upon its history in Scotland, and vividly describes the opposing influences :

" Plagiarist ! liar ! impostor ! heretic ! were among the expressions of malignant hatred lavished upon Galileo in 1609, as we learn from the record of the life of that eminent philosopher.

" The professor of philosophy of Padua refused to look through Galileo's telescope to see whether the satellites of Jupiter really existed, and he demonstrated to his own satisfaction that the facts could not be facts. In writing to Kepler regarding this, Galileo says : ' Oh, my dear Kepler, how I wish that we could have one hearty laugh together. Here, at Padua, is the principal professor of philosophy, whom I have repeatedly and urgently requested to look at the moon and planets through my glass, which he pertinaciously refuses to do.

Why are you not here? What shouts of laughter we should have at this glorious folly! and to hear the philosopher of Pisa laboring before the Grand Duke with *logical arguments*, as if with magical incantations to draw the new planets out of the sky.'"

The immediate reward which Harvey received, upon promulgating the doctrine of the circulation of the blood, was general ridicule and abuse, and a great diminution of his practice; and we are told by Hume that no physician in Europe, who at the time had reached forty years of age, ever, to the end of his life, adopted the doctrine of the circulation of the blood.

Sydenham, another eminent physician, whose improvements form an era in the history of medicine, was by many of his contemporaries called a quack and a murderer.

The author of "Fallacies of the Faculty" writes as follows:

"When a limb is amputated, the surgeons, to prevent their patients from bleeding to death, as you all know, tie the arteries. In the time of Francis the First they followed another fashion; then, and formerly, they were in the habit of stanching the blood by the application of boiling pitch to the surface of the stump. Ambrose Paté, principal surgeon to that king, introduced the ligature as a substitute; he first tied the arteries. Mark the reward of Ambrose Paté: he was hooted and howled down by the faculty of physic, who ridiculed the idea of hanging human life upon a *thread*, when boiling pitch had stood the test of centuries. In vain he pleaded the agony of the old application; in vain he showed the success of the ligature. Corporations, colleges, or coteries of whatsoever kind, seldom forgive merit in an adversary; they continued to persecute him with the most remorseless rancor; luckily he had a spirit to despise and a master to protect him against all the efforts of their malice.

"What physician now-a-days would dispute the value of antimony as a medicine? Yet when first introduced its employment was voted a *crime*. But was there no reason? Yes, it was introduced by Paracelsus—Paracelsus, the arch-enemy of the established practice! At the instigation of the college, the French Parliament accordingly passed an act making it penal to prescribe it.

"To the Jesuits of Peru, Protestant England owes the invaluable bark; how did Protestant England first receive this gift of the Jesuits? Being a " Popish remedy " they at once rejected the drug as the invention of the father of all Papists—the Devil !

"In 1693 Dr. Groenvelt discovered the curative power of cantharides, in dropsy ; what an excellent thing for Dr. Groenvelt! Excellent, indeed! for no sooner did his cures begin to make a noise than he was at once committed to Newgate prison, by warrant of the President of the College of Physicians for prescribing cantharides internally !. Blush, most sapient College of Physicians!—your

actual President, Sir Henry Halford, is an humble
imitator of the ruined Groenvelt."

"Lady Mary Wortley Montague, while abroad
with her husband in Turkey, had become acquainted
with the practice of inoculation for small-pox, and
on returning to England, in 1718, she attempted to
introduce it into the country.

" With indomitable courage she tried the experiment upon her own children, and was in consequence represented as an unnatural mother, who
cared nothing for her offspring. Lord Wharncliffe,
in his life of Lady Mary, tells us that "the faculty
all rose in arms, to a man, foretelling failure, and
the most disastrous consequences; the clergy
descanted from their pulpits on the impiety of thus
seeking to take events out of the hands of Providence, and the common people were taught to hoot
at her.

" We now read in grave medical biographies
that the discovery was instantly hailed, and the
method adopted by the principal members of that
profession. Very likely they left this recorded ; for
whenever an invention or a project (and the same
may be said of persons) has made its way so well,
by itself, as to establish a certain reputation, most
people are sure to find out that they always patron-
ized it from the beginning, and a happy gift of
forgetfulness enables many to believe their own
assertion.

"But what said Lady Mary of the actual fact
and actual time? Why, that the four great physi-

cians deputed by Government to watch the progress of her daughter's innoculation betrayed not
only such incredulity as to its success, but such an
unwillingness to have it succeed, such an evident
spirit of rancor and malignity, that she never cared
to leave the child alone with them, lest it should, in
some secret way, suffer from their interference."

At a later period, when Jenner was endeavoring
to introduce the process of vaccination, he was
assailed with the utmost ridicule by the members
of the learned profession of medicine.

Certain members of the clerical body discovered
vaccination to be "Antichrist," and the pulpit was
the vehicle for fulminations against it, in the same
manner as at a previous period against the innoculation of small-pox.

Dr. Chalmers, in speaking of the first reception
of the Newtonian philosophy, says: "Authority
scowled upon it, and taste was disgusted by it, and
fashion was ashamed of it." For *more than thirty
years* after the publication of Newton's discoveries, says Professor Playfair, the Cartesian system kept its ground, and actually the Newtonian
philosophy first entered the University of Cambridge under the protection of the Cartesian, by a
stratagem of Dr. Samuel Clark, who quietly explained the views of Newton, without any appearance of argument or controversy, in the form of
notes to a new translation which he published of
the French Cartesian work, long established as the
text-book by the tutors of the University.

With about the same amiable weakness, medical men of standing will admit the facts of magnetism under the names of Hypnotism, Braidism, &c., while perfectly oblivious of their existence under the names of Animal Magnetism, Mesmerism, &c., as of old. I sympathize with the wish to free the science from ill-assorted names; but the attempt to ignore the existence of all that has been said and done under these well-recognized appellations is quite as weak as the dodging of the Cambridge professors on the Newtonian philosophy.

When the proposal was made for the introduction of gas light, Sir Walter Scott ridiculed the idea, and in a letter to a friend, sneered at the folly of those who were actually talking of sending light through the streets *in pipes.* Sir Walter, however, had too much good sense to deny the existence of the light when it was actually produced; and besides becoming the chairman of a gas company in Edinburgh, he took advantage of its illuminating power at his residence at Abbotsford.

" Wallaston, the well known scientist, is said to have declared of a similar proposal that they ' might as well attempt to light London with a slice from the moon.' "

Dr. Elliottson states, in his *Human Physiology,* that when Laennec first published his great work he (Dr. E.) procured a stethoscope and investigated his statements. " For a length of time," he goes on to say, " I found some at the St. Thomas's Hospital treat percussion and auscultation with ridicule, some with absolute indignation, and others, for years, treated it with silent contempt, who all, I am happy to say, now practice both. I was, therefore, in the habit of studying them in the wards alone, and at hours when I expected to be unobserved. When at length I advocated and *taught* them in the school, one of my colleagues, I heard, pronounced it nonsense, or worse, in his lecture; and at the College of Physicians I heard a senior fellow, in a Croonian lecture, denounce the folly of carrying a piece of wood (some called the stethoscope *inutile lignum*) into the sick room, and making observations, to the destruction of all philosophical and dignified views, such as became men whose minds have been enlarged by the education which Oxford and Cambridge afford.

" When another fellow of the College was asked his opinion of auscultation in the wards of his hospital, he at once, as I am informed by the gentleman who asked the question, condemned it as nonsense, and when told that 'Elliottson assured his friends that he had a high opinion of it, and made his diagnosis of affections of the chest with infinitely more accuracy, by its means,' he replied: 'Oh! its just the thing for Elliottson to rave about!'

" Yet, good sense and truth have prevailed. This physician is now addressed as one who had the candor to examine auscultation at an early period, when others despised it, and who materially aided to spread its adoption."

The same eminent authority states that, for years after he published his work on prussic acid, in 1820, very few persons would employ it; and he was not only ill spoken of for recommending what was useless, but, till very lately, condemned for using dangerous poisons!

In 1824, the formula for prussic acid was withdrawn from the new edition of the Pharmacopœia, then in course of preparation; "yet," adds Dr. Elliottson, "it is now employed universally and daily by good practitioners of all ranks."

We have thus the melancholy fact demonstrated, that many of the greatest discoveries ever made were received at the outset with ridicule and contempt. Mesmerism endured precisely the same ungenerous treatment when put forward by Dr. Elliottson, to which allusion has already been made.

"It appears," says Mr. Colquhoun, "that there are some persons, even of note, members of learned incorporations, fellows of royal, and other privileged societies, professors in ancient universities, &c., to whom, at a certain period of life, the prospect of an accession of real knowledge, instead of being agreeable and satisfactory, is, on the contrary, rather unpleasant, painful, and humiliating.

"Every man who then ventures to present them with novel facts or ideas, or in any way attempts to rectify or extend their notions of things, is regarded by them as an invader,—a robber,—an enemy to what they have been accustomed to conceive to be their vested rights in literature and science. Goethe, the celebrated German poet, is reputed to have said, upon some particular occasion, that when, from time to time, a man arises who is fortunate enough to discover one of the grand secrets of nature, ten others immediately start up, who industriously and strenuously endeavor to conceal it again from view.

"It is so—was—and probably ever shall be. The conflict between light and darkness appears to be interminable.

"The race of the *obscurantist* in politics, in science, and in literature, promises to survive to the end of time. To use the language of a favorite old author, they are exceedingly angry with every one that hath outgrown his cherrystones and rattles, speak evil at a venture of things they know not, and, like mastiffs, are fiercer for being kept in the dark."

Lang observes:—"The doctors refused to look through Galileo's telescope; and because certain things were written in their books, they declined to examine the great book of nature for themselves. The old practitioners are resolutely opposed to innovation, and the more youthful, afraid of the frowns of their seniors, follow servilely in their footsteps. It was said that Dr. Elliottson for a time lost a large part of his practice, in consequence of his adoption and advocacy of magnetism, but this was only temporary, and it returned to him again with largely increased volume.

An able literary and political journal of the day, the *Examiner*, in noticing these attacks on Dr. Elliottson, says:—" If, as we apprehend to be the case, the existence of certain phenomena, undoubtedly of great interest and probably of great importance in a physiological view, is pretty generally admitted to be the result of recent experiments, it is high time to cease calling names, and begin rational discussion. The treatment to which Dr. Elliottson has been exposed from the time these questions were started, the members of a liberal calling should surely have reserved for the interested quack, or the vain pretender. There has been as little of either in the career of this distinguished physician, as in that of the foremost member of the profession he had so long assisted and adorned. Policy and worldly considerations apart—no man had better claims to be respectfully listened to. His admitted learning, his foregone recognized discoveries in medicine, his unimpeached veracity and high character, as they qualified him for that course which only the *few* are at any time fit to take, should have saved him from those vulgar imputations which the *many* are at all times prone to indulge.

" It is surely time that the word of an intelligent physician or surgeon—of a man whom the world would believe, without hesitation, on any ordinary topic—should at once be received when he unfolds truths of grave import to society. It is surely time to abandon implicit confidence in certain dogmas to be found in books, and to walk abroad and behold 'the visible and living world.'"

"There are few things," says Bailey, "more disgusting to an enlightened mind than to see a number of men, a mob, whether learned or illiterate, who have never scrutinized the foundation of their opinions, assailing with contumely an individual who, after the labor of research and reflection, has adopted different sentiments from theirs, and pluming themselves on the notion of superior 'knowledge,' because their understandings have been tenacious of prejudice." And Jobald, a French writer, remarks: "Galileo, Newton, Salomon de Caus, Volta, Fulton, Winser, Arkwright, Gall, and all who have presented themselves, with a truth in their hand, at the door of this great Bedlam called the world, have been received with stones and hisses."

Another author has said: "Whilst the unlearned were all busied in getting down to the bottom of the well where Truth keeps her little court, were the learned, in their way, as busy in pumping her up through the conduits of dialectic induction; they concerned themselves not with *Facts*—they *Reasoned.*"

The practical lesson to be learned from all this is to avoid the pitfalls which intolerant opinion has so often in the past digged for its own destruction, and learn to be careful of dogmatic assertion upon new discoveries that we have not personally investigated.

"Nothing," says Sir Humphrey Davy, "has so much checked the progress of philosophy as the confidence of teachers in delivering dogmas as truths which it would be presumptuous to question. It was this spirit which, for more than ten centuries, made the crude physics of Aristotle the natural philosophy of the whole of Europe. It was this spirit which produced the imprisonment of the elder Bacon and the recantation of Galileo. It *is* this spirit, notwithstanding the example of the second Bacon, assisted by his reproof, his genius and his influence, which has, even in later times, attached men to imaginary systems—to mere abstracted combinations of words—rather than to the *living* and *visible* world, and which has often induced them to delight more in brilliant dreams than in *beautiful and grand realities.*"

Fourcroy, the celebrated chemist, thus describes the difficulties which environed magnetism in his day: " The cold inactivity of some; the affected indifference of others; the contempt expressed by one person; the irritated self-love, and the languid attachment of another for the doctrine of his ancestors; the dread of novelty, and the prejudices of every kind; all the mean passions which, gliding into society and playing their parts in civil life, are also to be met with in the sciences; the sarcasms and epigrams with which they arm conversation— all these retard for a short time, perhaps for some years, the progress of new ideas; but truth ultimately overcomes every obstacle. Neither the

clamors of envy, nor the resistance of prejudice
nor the opposition of ignorance, can terrify it. It
is the rock against which the impotent waves of
human passion are broken.

"When the vivid light of truth strikes those
minds that are properly adapted to feel its influence, it soon inspires them with a sufficient degree
of force to make them proclaim it with confidence,
and to establish its rights on a solid foundation."

I would not argue that all these bitter words
hold good in regard to the reception of magnetism
in this day; still, they are true as against many
who close their eyes with prejudiced blindness even
now, else the words of Dr. Wm. B. Carpenter
would not have been penned in 1875, when he said,
in referring to mesmeric phenomena : "Some there
are who persist in the determination to disbelieve in
the genuineness of *all* the asserted facts, desig-
nating them as ' all humbug,' and maintaining that
none but fools or knaves could uphold such nonsense.

"Such persons, however, must now find themselves in the unenviable predicament of being
obliged to place some of their best friends in one
or other of these two categories ; since it is impossible to go into any kind of society, literary or scientific, professional or lay, gentle or simple, without finding a large proportion of intelligent and
truthful persons, such as would be regarded as
trustworthy on all other subjects, who affirm that
they have been themselves the actors in some one or

another of the performances in question, and that, however strange the phenomena may seem, they are nevertheless genuine." "The members of the medical profession, accustomed to the vagaries of hysteria, and recognizing the hysterical constitution of a large proportion of the subjects of mesmeric and spiritualistic agency, have too generally satisfied themselves with the phrase 'all hysterical,' a reply which affords no real information to those inquirers who think that their doctors ought to help them to a solution of such difficulties, and which has now been fully proved to be incorrect by the fact that steady, sensible, middle-aged men, having all their wits about them, are sometimes found to be as good subjects of certain of these operations as the susceptible young females who are deservedly regarded with so much suspicion." (Carpenter's Mental Physiology, p. 612.)

So it is quite evident that so eminent an authority of the present day as Dr. Carpenter feels called upon to give some of his professional brethren a cutting reproof for their stubborn unbelief and supercilious attitude.

The advance of the science and practice of magnetism has been delayed and obstructed by two classes of the community, one of which will steadily shut its eyes to ALL EFFECTS, no matter how well the facts may be authenticated, unless you are ready to furnish a full *explanation of cause simultaneously with the effect.* We have already seen how unreasonable and foolish such hard-headed, conservative wiseacres can become upon the presentation of anything new, that is opposed to their former and preconceived ideas, no matter what may be its promise of progress and utility. It is quite human and natural to demand such explanations, and it would be very gratifying if we could make them ; but, so far as I can see, little progress has been made in that direction, and I believe that more real progress *would* be made by relegating such questions for the present to the domain occupied by other incomprehensible *natural* forces which, like magnetism, are only known through THEIR effects.

Instead of multiplying doubtful hypotheses, which cannot satisfy this morbid curiosity, I would multiply instances of cure. Relief of human woe is a nobler work and a better answer to all cavilling.

The high tribute which Dr. Carpenter pays to the value of Hypnotic treatment in disease, in his valuable work on MENTAL PHYSIOLOGY, corroborated by the cases which he gives in illustration, is worth more than all the chapters he devotes to the probable WAYS by which nature works, although so entertainingly written; for, while he will not succeed in convincing men of the truth of *speculations*, opposed as they are in many cases by their own personal experience, his statements of cure will not be doubted—certainly not by any who know the complete trustworthiness of the author.

Another class of obstructionists in magnetic science is made up of that credulous, superstitious element of humanity, who, while believing the evidence of their senses, are disposed to jump to the conclusion that such wondrous effects must be of a MIRACULOUS or *supernatural* character.

They ignore all natural law, and if they are the recipients of the power in any remarkable degree, they claim for themselves some *supernatural* or personal endowment.

This class disgust with their pretensions many an honest seeker who might otherwise be quite willing to accept natural effects, although admitted to be extraordinary, and beyond explanation. We have only to use this beneficent force of nature as we would any other of her grand provisions for man's welfare, content with the blessings it bestows, and grateful for a power to relieve human pain and suffering so wisely provided by Him who never errs in establishing natural law, although He may not vouchsafe to us its full explanation.

In conclusion, let me quote the words of Baron Von Reichenbach, who, in philosophizing upon the outlook of the branch of this subject embraced in *odic force*, says : " He who has not grudged the labor of making himself acquainted with these results, will share my conviction that every physiologist and every physician must not only notice, but become thoroughly master of them (magnetic truths) if he is not contented to remain in ignorance of a whole series of important truths.

"The physiologist will discover in sensitives a group of vital functions which have been hitherto overlooked, and the physician will find in OD a power without which he will not only not be able to give any assistance in numerous cases, but even to understand the symptoms he beholds, much less to have any mastery over them. I do not mean that every physician should be a magnetizer; far from it : but a physician who understands nothing about the terrestrial magnetism of position, lunar and solar od, the odic poles of man, the odic influence of baths, the partial pass or odic obstructions, or the mighty influence of od on spasms,— nay, who does not know in what direction to stand when he feels a patient's pulse, must certainly play a miserable part by the bedside of a sensitive. The present and future revelation of the laws of od, must cause almost a revolution in whole sections of therapeutics."

APPENDIX.

DR. ELLIOTSON'S ADJURATION TO THE PROFESSION, AND DR. ESDAILE'S REMARKABLE PROTEST TO THE AMERICAN CONGRESS — MAGNETISM THE FIRST "ANÆSTHETIC."

Much has already been presented to establish the claims and prove the merits of vital magnetism in an exceedingly wide range of diseases; the difficulty has been to reduce the number of typical cases within the limits of a work like this. Passing by the vast array of authenticated facts to be found in the magnetic literature of continental Europe, the earnest student is asked to look at the record of a single English work to which reference has already frequently been made, viz.: The *Zoist*, in the thirteen volumes of which will be found over four thousand pages of facts and statements, *properly authenticated*, contributed mainly by distinguished members of the medical profession of England. To facilitate the investigation, an *index* is appended, indicating the form of disease and the number of the volume in which it may be found. These cases are detailed with great professional

minuteness, and have mainly been reviewed by that
eminent physiologist, Dr. John Elliottson, who, at
the conclusion of his Harveian oration, delivered
to the College of Physicians in London, June 27th,
1846, appealed to his compeers, in the portion of
the address devoted to Vital Magnetism, in these
terms:

"It is the imperative, the solemn duty of the
profession anxiously and dispassionately to determine these points by experiment, each man for
himself. I have done so for ten years, and fearlessly declare that the phenomena, the prevention
of pain under surgical operations, the production of
repose and comfort in diseases, and the cure of
many, even after the failure of all ordinary means,
are true.

"In the name, therefore, of the love of truth, in
the name of the dignity of our profession, in the
name of the good of all mankind, I implore you
carefully to investigate this important subject."

In the same spirit I respectfully call attention to
the annexed index; also to this "protest" addressed
to the American Congress by Dr. Esdaile, the true
discoverer of painless surgery:

*Protest and Petition of James Esdaile, M. D., Surgeon H. E.
I. C. S., to the Members of the American Congress :*

Respectfully showeth,—that the writer, a British subject,
and a surgeon in the East India Company service, has read .
with great astonishment the following statement in a Report
of the Select Committee appointed by the United States

Government, to decide who discovered the anæsthetic virtues of ether, so that Congress might bestow upon him the reward of $100,000.

The Report says : " At various periods, and in various ages, hope has been excited in the human breast that this great agent (the means of producing insensibility to pain) had been found; but all proved delusive, and hope as often died away, until the discovery now under consideration burst upon the world from our own country and in our own day.

" *Then, and not till then, was the time-cherished hope realized, that the knife would lose its sting, and that blood might follow its edge without pain.*"

In defense of truth and justice, I must take the liberty to inform Congress, that this statement is not only incorrect, but is perfectly untrue, and that your Committee have been grievously misled by culpably ignorant or corrupt witnesses,—corrupt to the extent of suppressing important evidence opposed to what is stated in the Report. In support of this assertion, I have to respectfully represent to Congress, that in April, 1845, having succeeded on the first trial in performing a painless operation upon a person in the *mesmeric* trance, I prosecuted the subject still further, and with such success that in eight months I had performed *seventy-three* PAIN-LESS OPERATIONS, many of them of the gravest description, an account of which was published by Messrs. Longman & Co., London, in 1846. This work is called" Mesmerism in India," and was im-

mediately reprinted in America. In July, 1846, I
reported to the Government of India that I had
performed upwards of one hundred painless operations, and offered to satisfy any number of persons, in whom the Government had confidence, of
the truth of my statements.

The Government hereupon appointed a committee to report upon mesmeric operations to be performed by me in their presence. Six severe operations were performed by me in the presence of
the Committee during a fortnight that they sat,
and the Committee arrived at the following general
result on the question of pain during the mesmeric surgical operations witnessed by them :

"That, in three severe cases, there is no proof
whatever that any pain was suffered, and that in
the three other cases the manifestations of pain
during the operation are opposed by the positive
statement of the patient that no pain was experienced."

The Government, in remarking upon the Report
of the Committee to the Chairman, said :

"So far has the possibility of rendering the most
serious surgical operations painless to the subject
of them been, in his Honor's opinion, established
by the late experiments, performed under the eye
of a committee appointed for the purpose, as to
render it incumbent upon the Government to afford to the meritorious and zealous officer, by
whom the subject was first brought to its notice,

such assistance as may facilitate his investigations and enable him to prosecute his interesting experiments, under the most favorable and promising circumstances. With this view, his Honor has determined, with the sanction of the Supreme Government, to place Dr. Esdaile for one year in charge of a small experimental hospital in some favorable situation in Calcutta, in order that he may, as recommended by the Committee, extend his investigations to the applicability of this alleged agency to all descriptions of cases, medical as well as surgical, and all classes of patients, European as well as native."

During the year of experiment, forty capital mesmeric operations were performed in the Mesmeric Hospital, and at the end of the year my success was acknowledged in a minute of the Governor-General in Council, and I was rewarded in being made a presidency surgeon of Calcutta, that the natives might continue to enjoy the benefits of Mesmerism.

So that painless surgery, by means of Mesmerism, had been reduced by me to a regular, every-day system, in Bengal, long before ether had been heard of. Of which fact Congress may still further satisfy themselves by referring to Dr. Huffnagle, Consul for the United States at Calcutta, who, in October, 1846, assisted at the removal of a scrotal tumor weighing 103 lbs., of which the patient was quite unconscious, and ultimately perfectly recovered.

The simple and notorious fact is, that painless surgery by means of Mesmerism, years before ether was heard of, was as common in my hospitals as it has since become in Europe under the influence of chloroform, and nearly three hundred capital mesmeric operations had been performed by me before leaving India, two years ago.

This being the actual state of the case regarding the history of painless surgery, the American Congress will perceive that I am fully justified in asserting that their committee have been misled by ignorant or corrupt witnesses; and that however deserving of reward the discoverer of ether may be, he was only copying by a drug what had already been done by myself and others, by a much safer and pleasanter natural power mercifully implanted in the human constitution.

Congress will, I doubt not, be indignant at the daring and unprincipled suppression of evidence that I have exposed to them, and will be at a loss to imagine the motives for such an offense.

As politicians, the members of Congress must be aware of the perverting, distorting, unfair representations of party spirit, by which history itself has become "*a lie,*" according to a celebrated English statesman.

Unfortunately this vice is not confined to politics, but exists in all professions, and in an intensely disgraceful degree, I lament to say, in that of medicine.

Mesmerism, from its truth, has been excommunicated by the doctors without knowledge or examination, and all the medical journals having rashly and ignorantly pronounced against it, afterward conspired together in defense of their wrong-doing, and trusted to extinguish the obnoxious doctrine by keeping the medical profession and the public in total ignorance of the matter. They have hitherto succeeded so well in this that it is possible the medical witnesses examined by your committee may only have been laboring under the most gross ignorance in giving their evidence ; but the American Congress will not lower itself so much in the eyes of the civilized world as to adopt the verdict of an ignorant and incompetent jury.

If, on the other hand, it shall appear that important evidence has been deliberately suppressed by the witnesses, then the world will expect that Congress, in its wisdom and justice, will repudiate the report of its committee, and severely punish those who would have misled it.

It is with no intention of competing for the *dollars* in question that I have taken the liberty to address you (although the man who proved that painless surgery could be practiced on a whole people, might look for some reward), as these seem irrevocably destined to "one of three citizens of the United States" who had the sense to seek for some drug by which to imitate what I had already done by Mesmerism.

But, in the new judicial inquiry recommended by the Committee, to determine which of my three imitators is to be the lucky winner of the dollars,

I respectfully petition Congress to be permitted to adduce proof in support of what I have now advanced, by which an honorable opportunity will be afforded for correcting the errors in the Committee's Report, and Congress will be saved the shame of sanctioning by its high authority another of the lies of history.

I have the honor to be,

Your most obedient servant,

JAMES ESDAILE, M. D

FAIRMOUNT, Perth, Scotland,
8th August, 1853.

Says Dr. John Elliottson, in reviewing Dr. Esdaile's practice, in 1853:

" These operations consisted of some of the most appalling in the annals of surgery, and we speak practically when we say, that we do not believe a similar result will be obtained, as regards the absence of fatal cases, by the inhalation of ether or chloroform. The one is a *health-restoring power*— the other is a drug; one enables the nervous system to sustain great shocks without leaving any injurious consequences. The other produces a state of profound stupor, requiring great care during its exhibition, and even, with the greatest care, frequently fatal."

" The rapid production of mesmeric anæsthesia is the great desideratum of the day "

LITERATURE OF VITAL MAGNETISM.

Animal Magnetism. By Alphonse Teste, M. D. D. Spellman, M. D., A. M., Translator. Fellow of the Dublin College of Physicians. London, 1843.

Animal Magnetism. By John Ashburner, M. D. Member of the Royal Irish Academy. London, 1867.

Mesmerism in India and its Practical Application in Surgery and Medicine. By James Esdaile, M. D. Presidency Surgeon, Bengal, India. 1843. Reprint by Silas Andrews & Son, Hartford, 1850.

Human Magnetism. By W. Newnham, M. R. S. L. The Utility of its Application for the Relief of Human Suffering. London, 1845.

Hypnotism (Neurhypnology), or the Rationale of Nervous Sleep ; with numerous cases of its successful application in the relief and cure of diseases. By James Braid, M. D., Surgeon and M. R. C. S. E., C. M. W. S., &c. London, 1848.

Facts in Mesmerism. By Rev. Chauncy Hare Townshend, A. M. Late of Trinity Hall, Cambridge. London, 1844.

Researches on Magnetism, Electricity, Heat, Light, Crystallization and Chemical Attraction in their relations to Vital Force. By Karl, Baron Von Reichenbach, Ph. Dr. Translated by Wm. Gregory, M. D., F. R. S. E. Professor of Chemistry in the University of Edinburgh. London, 1850.

Isis Revelata: An Inquiry into the Origin, Progress and Present State of Animal Magnetism. By J. C. Colquhoun. Advocate. Edinburgh. 1844.

Report of the Experiments on Animal Magnetism made by a Committee of the Medical Section of the French Royal Academy of Sciences. Read at the meetings of the 21st and 28th of June, 1831. Translated by J. C. Colquhoun. Edinburgh, 1833.

Mesmerism and its Opponents. By Rev. J. Sandby, M. A. London, 1848.

History, Phenomena and Practice of Mesmerism; with reports of Cases Developed in Scotland. By William Lang. 1843.

Curative Powers of Mesmerism; with one hundred and fifty cases covering various forms of disease. By Thomas Capern. London, 1851.

Mesmerism Not Miracle. By a Mesmeric Practitioner. Showing that Mesmeric Cures are not Supernatural. By Geo. Barth. London, 1853.

Letters on Mesmerism. By Harriet Martineau. London, 1845.

Somnolism and Psycheism, Phenomena of Nervation as Revealed by Vital Magnetism or Mesmerism, considered Physiologically and Philosophically; with Mesmeric Experience. By Joseph W. Haddock, M. D. London, 1851.

Seven Lectures on Somnambulism. Translated from the German of Dr. Arnold Wienholt. By J. C. Colquhoun. Edinburgh, 1845.

Animal Magnetism Delineated by its Professors: a Review of its History in Germany, France, and England. Report from British and Foreign M. R., 1839.

Annales du Magnetisme. 8 volumes. Paris, 1814–16.

Bibliotheque du Magnetisme Animal. 8 volumes, *Ib.* 1817–19.

Healing by the Hand and Will. By Sir Charles Isham. London, 1862.

Elemens du Magnetisme Animal. M. De Lausanne. Paris, 1818.

Puissance de L'Electricité Animal, ou Magnetisme Vital. Dr. J. Pigeaire. Paris, 1839.

Lectures on Zoistic Magnetism. By W. Luresby, D. D., F. R. S. London, 1849.

Mesmerism in Disease. Plain Facts and Cases. By Dr. H. Storer. London, 1845.

Magnétisme et Magnetotherapie et sur Le Gyro Magnetisme. By Le Compte de Szapary. Paris, 1854.

Manual Practique de Magnetisme Animal et son Application au Traitement des Maladies.

Animal Magnetism : History of the Principles ; with Cures and Cases. By Dr. G. Winter, 1801.

Vital Magnetism. By Rev. T. Pyne. London, 1844.

Report of Cases treated in the Mesmeric Hospital, from June to December, 1847. With the Reports of the official Visitors. Printed by order of Government. December, 1848.

Philosophy of Animated Nature ; or the Laws and Action of the Nervous System. By G. Calvert Holland, M. D. ; Physician Extraordinary to the Sheffield General Infirmary. London, 1848.

The Journal of Psychological Medicine and Mental Pathology. Edited by Forbes Winslow, M.D., England, 1849.

Illustrations and Enquiries relating to Mesmerism. By the Rev. S. R. Maitland, D. D., F. R. S., F. S, A. London, 1849.

The Hand Book of Mesmerism. For the Guidance and Instruction of All Persons who desire to practice Mesmerism for the cure of Diseases. To which is annexed the Rules and Regulations of the Mesmeric Infirmary, No. 9 Bedford St., Bedford Square, London ; with a list of the Subscribers to it. 1850.

Practical Instruction in Animal Magnetism. By J. P. F. Deleuze. Translated by T. C. Hartshorn. With Notes, and a Life, by Dr. Foissac. London, 1850.

The Principle of Health Transferable. (Two Editions.) London, 1850.

Mesmerism : Its Processes, Uses and Advantages Explained ; with Directions for its Application in the Sick Chamber. By S. D. Saunders, late Hon. Secretary to the Bristol Mesmeric Institute. London, 1850.

The Mighty Curative Powers of Mesmerism ; with a great mass of cures. By Thomas Capern, Secretary to the London Mesmeric Infirmary. London, 1851.

Letters to a Candid Enquirer on Animal Magnetism. By William Gregory, M. D., F. R. S. E., Professor of Chemistry in the University of Edinburgh. London, 1851.

Lectures on Mesmerism. Delivered at the Rotunda, Dublin. By J. W. Jackson. 1851.

" We have heard much of *contagion,* and the word is one of terror ; but it would seem that nature has been more just than we deemed, for the principle of health is as transferable as that of sickness, and a vigorous operator is a fountain of sanative influence whence the decrepit and ailing may derive a fresh supply of vital force."

Recueil d'Operations Chirurgicales Practiquées sur les sujets Magnétisés. Par A. Loysel, Docteur en Medicine, á Cherbourg. Cherbourg, 1852.

The Magnetoscope. A Philosophical Essay on the Magnetoid characteristics of Elementary Principles, and their relations to the Organization of Man. By T. Leger, Doctor of the Medical Faculty of Paris ; late Professor of Anatomy at the Practical School ; Fellow of the Society of Sciences and Arts of the Department of the Marne ; late Professor of the Medical College of Mexico, etc., etc. London, 1852.

Buckland's Hand Book of Mesmerism. (Three editions.) London, 1852.

Mesmerism Solved, Divested of Mystery, and shown to be simply an overlooked Branch of Medical Science, to be received and practiced as an auxiliary by the Medical Professors and Heads of Families, for curative purposes. By Dr. Jones. London, 1853.

Hypnotic Therapeutics; illustrated by Cases. By James Braid, M.R.C.S., Edin., M.U.S., etc. Reprint from the Monthly Jouinal of Medical Science. 1853.

Mesmerism, and the Diseases to which it is most applicable. By Falconer Mills. Dublin, 1854.

Medicina Mentis. By Hugh Hastings, M.D. Cheltenham, England, 1854.

The Illustrated Practical Mesmerist, curative and scientific. By William Davey. Edinburgh, 1854.

Bulletin de l'Athénée Magnetique de Lyon. Journal des Sciènces Psycho-Physiques. Lyons, 1854.

Errors Dispelled; or, Mesmerism without Sleep, and Mesmerism with Medicine. By S. D. Saunders, Medical Mesmerist. Clifton, England, 1855.

Heilkraft des theirischen Magnetismus nach eigenew Beobachtungen. By Dr. Weinholt, of Bremen. 3 vols. 8vo : 1802–1805.

Le Magnetisme Animal, considéré comme moy en Thérapeutique ; son application au Traitment de deux cas remarquables de Neuropathic Par Charles de Résimont, Docteur en Medicine de la Faculté de la Paris, 1843.

Journal du Magnetisme. Par une Societé de Magnetiseurs et des Médecins sous la direction de M. Le Baron Du Potet. 1845.

The Curative Power of Vital Magnetism verified by Actual Application to numerous cases of disease. By Dr. Jones. London, 1845.

The Harveian Oration delivered before the Royal College of Physicians, London. By Dr. John Eliottson, Cantab. F.R.S., 1846.

Should not investigation take a practical turn in this country when we look abroad, see how many able minds have grappled successfully with the problems of this branch of vital science, and demonstrated its possibilities?

These are but a tithe of the whole, but sufficient, I hope, to stimulate and instruct the student in this branch of physics.

INDEX TO CASES IN THE "ZOIST."

Abscess	. . . . .	Volume 13
Abscess, Lumbar	. . . . .	6, 9
Acid, nitric, painless application of	. .	7
After pains	. . . . .	5, 11, 13
Ancles, weakness of	. . . . .	10
Ancles, sprained	. . . . .	11, 13
Arm, stiff	. . . .	6
Asthma	. . . . .	1, 2, 3, 4, 7, 9, 15
Bladder	. . . . . .	4, 5, 8
Blindness	. . . .	3, 7, 8, 9, 10, 11, 12
Bone cured	. . . .	12
Brain, overworked	. . . . . .	3
Brain, inflammatory excitement of	. .	6, 7,
Breasts, inflamed	. . . . .	11, 12
Bronchitis	. . . .	3
Burns	. . . .	10, 12
Catalepsy	. . . .	13
Caustic, painless application of	. . .	4
Chest	. . . .	5, 12
Chilblains	. . . .	11
Chlorosis	. . . .	8
Cholera	. . . . - . .	9

Coma cured 11
Constipation 1
Consumption 2, 3, 8, 12
Contractions 3, 6
Convulsions 4, 11. 12
Cough 6, 10
Cutaneous diseases 10, 11, 12
Deafness 2, 4, 5, 8, 11, 12, 13
D. B. B. Morrison, also reporting 211 cases of
    deafness cured by him 12
Debility 3, 5, 7, 9, 10, 12, 13
Delirium 2, 12
Delirum tremens 8, 11, 13
Diarrhœa 11, 13
Dropsy 7, 8, 10, 12
Dyspepsia 5, 12, 13
Ear abscess of 8
Earache 11
Elephantiasis 11
Epilepsy . . 1, 2, 3, 4, 5, 6, 7, 8, 9, 10, 11, 13
Erysipelas 7, 8, 10, 12
Eyes, agony of 7
Eyelids, spasm of 10
Emaciation. 8
Face, swelled 6
Female ailments 4
Fever 7, 13
Fingers 9, 12, 13
Foot 12
Fractures 9, 11
Ganglion 12

Gout 5, 6, 8, 9, 10, 12
Gout, rheumatic . '. 11
Glands, enlarged 6
Grief, effects of, cured 1
Gums 7, 11
Hand 11, 13
Headache . . 2, 3, 4, 5, 6, 7, 8, 9, 10, 11, 12, 13
Heart, affection of . . . 1, 3, 5, 6, 8, 12
Hiccough 2, 5
Hip 9
Hœmorrhage 7, 8
Hydrophobia 3, 6, 9
Hypochondriasis . . . 2, 3, 4, 5, 7, 10, 11, 13
Hysteria 1, 3, 5, 6, 8, 9, 10, 11
Inflammation 3, 4, 10, 11, 12
Insanity . . . 3, 4, 5, 6, 7, 8, 9, 10, 11, 12
Joints, enlarged 10
Jumping, fit 6
Kidney 9
Knee, diseased . . . 2, 5, 6, 8, 10, 11, 12
Leg, diseased 8, 12
Leprosy 12
Liver, diseased 12, 8
Lockjaw 6, 7, 10, 12, 13
Lung, diseased 4, 7, 11, 13
Lung, bleeding from 8
Muscular pains, cured 1
Neck, twisted 5, 8
Neck, rigid 5, 12, 13
Nervousness, cured . . 2, 3, 7, 8, 10, 12, 13
Neuralgia 1, 2, 3, 4, 5, 6, 7, 8, 9, 10, 11, 12, 13

Operations, painless surgical, 1, 2, 3, 4, 5, 6, 7, 8, 9, 10, 11, 12, 13.

Opthalmia, cured . . 2, 3, 7, 8, 9, 10, 11, 13

Ovarian, disease 12, 13

Pain, relieved . . . 4, 5, 6, 7, 9, 10, 11, 13

Palpitation, cured 2, 7, 8, 10

Parturition, painless . . 2, 4, 6, 8, 9, 10

Paralysis, cured. . 1, 4, 5, 6, 7, 8, 9, 10, 11, 12, 13

[Delueze also reports over 60 cases of cure in France.]

Pericarditis 9

Polypus 8, 9

Prolapsus, reduced 4

Prostate Gland 7

Quinsy 2, 7

Rheumatism, cured, 1, 2, 3, 4, 5, 6, 7, 8, 9, 10, 11, 12, 13.

Ringworm 12

St. Vitus's dance, cured . . 1, 3, 5, 6, 7, 10

Scrofula 3

Short sightedness 5

Sleeplessness 6

Speech, recovered . . 2, 4, 5, 6, 7, 8, 9, 10, 11

Spinal affection, cured . . 2, 4, 5, 7, 8, 9, 12

Stomach, pain of 5, 8

Tabes, mesenterica 13

Tetanus 5, 6

Throat disease 6, 12

Toothache 7, 10, 11, 12

Trance 5, 9, 13

Ulcers 3, 6, 7, 8, 9, 13
Unconscious action of brain . . . 6
Uterine disease 3, 8, 11
Vomiting 7, 8, 12

www.ingramcontent.com/pod-product-compliance
Lightning Source LLC
Chambersburg PA
CBHW021657210326
41599CB00013B/1447